Memories in Wireless Systems

Rino Micheloni · Giovanni Campardo ·
Piero Olivo (Eds.)

Memories in Wireless Systems

 Springer

Editors

Rino Micheloni
Qimonda Italy srl
Design Center Vimercate
Via Torri Bianche, 9
20059 Vimercate (MI)
Italy
rino.micheloni@qimonda.com

Giovanni Campardo
STMicroelectronics Srl
Memory Product Groups
Wireless Flash Division
Via C. Olivetti, 2
20041 Agrate Brianza MI
Italy
giovanni.campardo@numonyx.com

Piero Olivo
Universitá di Ferrara
Dipto. Ingegneria
44100 Ferrara
Italy
olivo@ing.unife.it

ISBN: 978-3-540-79077-8   e-ISBN: 978-3-540-79078-5

Library of Congress Control Number: 2008930841

Originally published in Italian "Memorie in sistemi wireless", Edizioni Franco Angeli, 2005,
ISBN 13: 978-88-4646-779-9

Cover design: WMXDesign GmbH

Printed on acid-free paper

9 8 7 6 5 4 3 2 1

springer.com

Original cover of the italian version. Laura and Greta Micheloni drawing

About the Editors

Rino Micheloni (Senior Member, IEEE) was born in San Marino in 1969. He received the Laurea degree (*cum laude*) in nuclear engineering from the Politecnico di Milano, Milan, Italy, in 1994. In the same year, he was with ITALTEST, Settimo Milanese, Italy, working on industrial non-destructive testing reliability. In 1995 he joined the Memory Product Group, STMicroelectronics, Agrate Brianza, Italy, where he worked on an 8-Mb 3-V-only Flash memory, especially on the analog circuitry of the read path. He was the project leader of a 64-Mb 4-level Flash memory and, after that, he designed a 0.13-μm test chip exploring architectural solutions for Flash memories storing more than 2 bits/cell. Then he was the Product Development Manager of the NOR Multilevel Flash products for code and data storage applications. From 2002 to 2006 he led the NAND Multilevel Flash activities and the Error Correction Code development team. At the end of 2006 he joined Qimonda Flash GmbH, Unterhaching, Germany, as Senior Principal for Flash Design. Currently, he is with Qimonda Italy srl, Vimercate, Italy, leading the design center activities.

He is the author/co-author of more than 20 papers in international journals or conferences. He is co-author of Chapter 6 in *"Floating Gate Devices: Operation and Compact Modeling"*, Kluwer Academic Publishers, 2004 and of Chapter 5 in *"Flash Memories"*, Kluwer Academic Publishers, 1999. He is co-author of the books *"VLSI-Design of Non-Volatile Memories"*, Springer-Verlag, 2005 and *"Memorie in sistemi wireless"*, Franco Angeli, 2005. Mr. Micheloni was co-guest editor for the Proceeding of the IEEE, April 2003, Special issue on Flash Memory. He is author/co-author of more than 100 patents (79 granted in USA).

In 2003 and 2004 he received the STMicroelectronics Exceptional Patent Awards for US patent 6,493,260 "Non-volatile memory device, having parts with different access time, reliability, and capacity" and US patent 6,532,171 "Nonvolatile semiconductor memory capable of selectively erasing a plurality of elemental memory units" respectively. In 2007 he received the Qimonda Award for IP impact.

Giovanni Campardo was born in Bergamo, Italy, in 1958. He received the Laurea degree in nuclear engineering from the Politecnico of Milan in 1984. In 1997 he graduated in physics from the Universita' Statale di Milano, Milan.

After a short experience in the field of laser in 1984, he joined in the VLSI division of SGS (now STMicroelectronics) Milan, where, as a project leader, he designed the family of EPROM nMOS devices (512k, 256k, 128k and 64k) and a Look-up-table-based EPROM FIR in CMOS technology.

From 1988 to 1992, after resigning from STMicroelectronics, he worked as an ASIC designer, realizing four devices. In 1992 he joined STMicroelectronics again, concentrating on Flash memory design for the microcontroller division, as a project leader. Here he has realized a Flash + SRAM memory device for automotive applications and two embedded Flash memories (256k and 1M) for ST10 microcontroller family. Since 1994 he has been responsible for Flash memory design inside the Memory Division of SGS-Thomson Microelectronics where he has realized two double-supply Flash memories (2M and 4M) and the single-supply 8M at 1.8V. He was the Design Manager for the 64M multilevel Flash project. Up to the end of 2001 he was the Product Development Manager for the Mass Storage Flash Devices in STMicroelectronics Flash Division realizing the 128M multilevel Flash and a test pattern to store more than 2 bits/cell. From 2002 to 2007, inside the ST Wireless Flash Division, he had the responsibility of building up a team to develop 3D Integration in the direction of System-in-Package solutions. Now he is responsible for activities of CARD Business Unit, inside the Numonyx DATA NAND Flash Group.

He is author/co-author of more than 100 patents (68 issued in USA) and some publications and co-author of the books *"Flash Memories"*, Kluwer Academic Publishers, 1999, and *"Floating Gate Devices: Operation and Compact Modeling"*, Kluwer Academic Publishers, January 2004. Author of the book *"Design of Non-Volatile Memory"*, Franco Angeli, 2000, and *"VLSI-Design of Non-Volatile Memories"*, Springer Series in ADVANCED MICROELECTRONICS, 2005, *"MEMORIE IN SISTEMI WIRELESS"*, Franco Angeli Editore, collana scientifica, serie di Informatica, 2005.

He was the co-chair for the "System-In-Package-Technologies" Panel discussion for the IEEE 2003 Non-Volatile Semiconductor Memory Workshop, 19th IEEE NVSMW, Monterey, CA. Mr. Campardo was the co-guest editor for the Proceeding of the IEEE, April 2003, Special issue on Flash Memory.

He was lecturer in the "Electronic Lab" course at the University Statale of Milan from 1996 to 1998. In 2003, 2004 and 2005 he was the recipient for the "ST Exceptional Patent Award", respectively: US patent 5,949,713 titled "Non volatile device having sectors of selectable size and number", US patent 6,493,260 titled "Nonvolatile memory device, having parts with different access time, reliability, and capacity", and US patent 6,532,171, titled "Nonvolatile semiconductor memory capable of selectively erasing a plurality of elemental memory units". Mr. Campardo is a member of IEEE.

Piero Olivo was born in Bologna (Italy) in 1956. He graduated in electronic engineering in 1980 at the University of Bologna, where he received the PhD degree in 1987. In 1983 he joined the Department of Electronics and Computer Systems of the University of Bologna where he became associate professor of Electronic Instrumentation and Measurements in 1992. In 1994 he became full professor of Applied Electronics at the University of Catania (Italy). In 1995 he joined the University of Ferrara (Italy) where, since 2007, he is dean of the Engineering Faculty.

In 1986–1987 he was a visiting scientist at the IBM T.J. Watson Research Center. The scientific activity concerned several theoretical and experimental aspects of microelectronics, with emphasis on physics and reliability of electron devices and non-volatile memories as well as design and testing of integrated circuits. In particular he is author of the first paper describing and analyzing stress-induced leakage current (SILC) in thin oxides and of the first analytical theory of aliasing error in signature analysis testing techniques.

Introduction

If the inhabitants of the Western world were asked to name the technological innovation that most changed their habits, they would respond in a number of ways. Depending on their age and financial means, the answers might range from modes of transportation (aeroplane, train, or automobile) to information systems (PC, computing devices in general), from household appliances (refrigerator, washing machine) to systems of communication (telephone, radio, television). Actually, all the inventions mentioned deeply influenced the lifestyle habits of the 20th century and have become an integral part of our daily life, to the extent that we would be incapable of giving any of them up.

Few would probably have answered that the most significant invention is the mobile telephone. However, if they were asked which invention had managed to infiltrate their lives most rapidly, the answer would be nearly unanimous: over a short number of years, the mobile telephone has transformed from a status symbol for a select few to a system of the most widespread use, one that has accelerated the process of penetrating the market and the habits of the population at large to an extent that other technological innovations achieved only over the span of decades (e.g., automobiles, household appliances, television sets, etc.).

The astoundingly rapid penetration of mobile telephony into the market and into everyday habits has been accompanied by an equally swift (and necessary) technological evolution in the mobile telephones themselves: the reduction in size, the development of new services, the standard of reliability demanded, the short period of time required for the design of new systems, and the competition among the few manufacturers who have managed to survive in a fierce battle to win market share—all have driven over recent years technological development in the fields of integrated circuits, semiconductor memories, application software, and so on.

By limiting our observations to the field of semiconductor memories, which have now reached the complexity of outright electronic macrosystems, we can see how the evolution of mobile telephones has led to the rapid diffusion of nonvolatile memories such as flash, i.e., memories that can maintain information even in the absence of a power supply. The traditional flash NOR memories, used to store the code utilized by the mobile phone for its functioning, have been flanked by the flash NAND memories, which are utilized for the storing of images, music, etc.

Chapter 1
Hardware Platforms for Third-Generation Mobile Terminals

D. Bertozzi and L. Benini

1.1 Introduction

The development of the Internet has allowed access to multimedia content for an increasing number of users. Desktop PCs or laptop computers have been the traditional access terminals to this kind of information. However, the evolution of technologies for wireless network connectivity (e.g., IEEE Wireless LAN 802.11 standards, GPRS, third-generation mobile telephony) has paved the way for high-speed Internet and multimedia content access even for portable devices such as handheld computers and media-enabled mobile phones.

Consider, for example, *i-mode*, the wireless Internet access service of DoCoMo (Japan's largest mobile carrier), which plans to make the functionality offered by its mobile terminals even more pervasive: it ranges from traditional mobile telephony services, e-messaging, gaming, and Internet access to more advanced services such as bar code reading, audio/video multimedia streaming, payments, self-certification documents, remote control of home appliances or of electromechanical devices, and the purchase of tickets [1]. In essence, this is an embodiment of the concept of ubiquitous computing, where networked electronic devices are integrated at all scales into everyday objects and activities to perform information processing [2].

The development of new technologies for network access and the extension of services made available to mobile terminals pose tight computation requirements to these latter devices, in order to support new functionalities and to improve the quality of offered services [3].

One representative case study is the evolution of standards for video sequence encoding. Since the introduction of the H.261 standard in 1990, more advanced standards have been introduced: MPEG-1 video (1993), MPEG-2 video (1994), H.263 (1995, 1997), and MPEG-4 visual (1998). This development was driven by advances in compression technology and by the need to adapt to new applications

D. Bertozzi
University of Ferrara, Ferrara, Italy

L. Benini
University of Bologna, Bologna, Italy

R. Micheloni et al. (eds.), *Memories in Wireless Systems*,
© Springer-Verlag Berlin Heidelberg 2008

and communication networks. Similarly, the recent emphasis on interactive applications, such as video-telephony on mobile communication networks (posing low latency constraints), and on noninteractive applications, such as standard definition TV streaming on error-prone channels (e.g., GSM and UMTS) or on best effort networks (e.g., the Internet), has led to the development of a new standard for efficient video coding: H.264/AVC [4]. H.264 allows a bit rate saving of 37% with respect to MPEG-4 visual for video streaming applications and of 27% with respect to H.263 CHC for video conference applications. However, the increase in coding efficiency is achieved at the price of a higher implementation cost. The encoder complexity is larger than that of MPEG-4 by more than one order of magnitude, while the decoder complexity is two times as much.

In this context, mobile terminals are evolving toward true multimedia platforms, from a hardware architecture viewpoint. On the one hand, scalable computation horsepower is achieved through the increased complexity of computation engines and their specialization for specific functions, but mainly through the integration of an increasing number of processing cores operating in parallel. On the other hand, these terminals must be able to support a large number of ever evolving telecommunication standards and their often conflicting requirements on system hardware resources.

Third-generation chipsets (such as [5]) are in fact designed to support data rates of several hundred kbps; high performance microprocessors with memory management units (e.g., ARM926J-S); hardware accelerators for Java applications; low power digital signal processors (DSPs); integrated hardware encoders and decoders for MPEG-4, JPEG, 2D/3D graphics engines, baseband Bluetooth processors, and wide band stereo codecs; advanced interfaces for LCD controllers; digital camcorders; complex memory hierarchies; GPS processors and modems for CDMA, GSM, GPRS, and UMTS; as well as band conversion circuitry and transmission power amplifiers. A key feature of modern platforms is power management support by means of dedicated devices such as voltage regulators and controllers for dynamically scaling the supply voltages of processing cores. More complex power management policies are often implemented, from those targeting specific hardware resources (e.g, backlight driver for LCD displays, battery discharge control) to system-wide power control policies (e.g., the definition of power states associated with different system functionality and power dissipation levels) [6].

Chipsets often come with adequate software support, in particular application programming interfaces (APIs) for fast access to system resources and for the development of embedded applications. In general, the software architecture of mobile platforms is structured in multiple software layers, the application and the system software representing the top layers of the stack. Multimedia applications usually consist of a collection of multiple media codecs with data dependencies. Provided proper standard interfaces are defined for these components, the application software can be reused across different platforms, regardless of the hardware or software implementation of the basic components or of their specific vendor. The integration layer of the OpenMax standard for media portability is an example thereof.

A lower software layer denoted hardware abstraction layer (HAL) is in charge of exposing an abstract representation of the underlying hardware platform, which can then be accessed through a set of properly defined HAL APIs. The abstraction gap between the application software and the HAL is usually bridged by a real-time operating system (RTOS) and by middleware libraries.

Finally, it is worth mentioning that the rapid development of parallel multimedia platforms for mobile terminals with wireless connectivity is pushed by the continuous scaling of silicon technology. The International Technology Roadmap for Semiconductors (ITRS) [7] confirms that the semiconductor industry is substantially following Moore's Law, which states that the number of transistors placed on an integrated circuit doubles approximately every two years. Microprocessors and ASICs were produced in 1998 in 250 nm technology, scaled down to 180 nm in the year 2000, 130 nm in 2002, and 90 nm in 2004. The advent of the 65 nm technology node in commercial products was expected in 2007. The shrinking of feature sizes translates into an increase of processor clock frequencies (and hence of performance) and in a reduction of supply voltages to keep power consumption under control. On the other hand, a number of issues arise associated with the reduction of noise margins, the increase of leakage power, and process variations.

A main aspect of the device scaling trend consists of the capability to integrate a large number not only of computation units, but also of I/O units, memory devices, and communication fabrics on the same silicon die. In essence, an entire system can now be integrated on the so-called *Systems-on-Chip (SoCs)*. Typical application domains include embedded systems for wireless networks or automotive components. SoCs allow cutting down on design time by means of the reuse of predesigned and preverified processing or I/O cores across different platforms, and most of the front-end design effort is now devoted to core integration into the system and not to core design [8].

As an example, the scalability limitations and the power inefficiency of state-of-the-art bus-based communication architectures are becoming evident as the integration densities increase [9]. This is requiring a fast evolution of communication protocols and topologies and even the advent of disruptive communication technologies.

This chapter focuses on the main components that build up state-of-the-art mobile multimedia platforms, with emphasis on third-generation mobile phones.

1.2 Computation Units

Processor cores required by mobile multimedia platforms must meet some fundamental requirements: support for operating systems typically found in embedded systems such as Symbian, WindowsCE, PalmOS, or Linux; computation efficiency for multimedia signals (audio, video, silent images, 2D and 3D graphics, games); and support for Java and for cryptography algorithms.

Several ways do exist to increase the computation efficiency of general purpose processors. Some of them aim at extending the instruction set architecture in the direction of DSPs while preserving full programmability. Other solutions progressively increase computation efficiency at the cost of programmability, such as DSPs, programmable coprocessors, and specialized hardware accelerators.

An overview of commercially available processor architectures employed in mobile multimedia platforms follows [10]. Some processors do not provide hardware extensions for digital signal processing. The lack of specific optimizations makes high frequency operation a key requirement for these processors in order to sustain execution performance. A typical example is represented by the Samsung *S3C24xx* series [11]: it is a monolithic processor architecture based on the ARM9 RISC processor core (Fig. 1.1a). The programming model is very simple, the architecture stable, and the roadmap well defined, with excellent support in terms of development toolchain and operating systems. This solution provides moderate performance at a low cost.

A different approach is taken by the *ARM9E* processors [12], based on minimal extensions of the instruction set architecture for signal processing. They are equipped with a multiplier, which allows the execution of multiply-accumulate (MAC) instructions in a single clock cycle, and with a reduced 16-bit instruction set (denoted *ARM THUMB*), which optimizes code density up to 25%. Some processors of the series (e.g., *ARM926EJ-S*) support dedicated hardware accelerators for Java (*ARM Jazelle* technology).

The Intel *PXA27x* takes an even more radical approach [13]. It consists of a SoC for wireless platforms based on the Intel *Xscale* microarchitecture and on the Intel *Wireless MMX* technology. Intel *Xscale* implements the *ARMv5TE* instruction set and includes a seven-stage pipeline. While the family of Intel *PXA255* processors provides a 40-bit accumulator and 16-bit SIMD operations for efficient audio/video decoding, the new *PXA27x* family extends the *Xscale* core with the *Wireless MMX* coprocessor and its instruction set (Fig. 1.1c). *Wireless MMX* provides an equivalent functionality with respect to the instructions of the *MMX* and *SSE* instruction sets previously used by Intel processors for desktop PCs such as the Pentium 4.

For this reason, the programming style is well known to software developers. The implemented SIMD instructions are 64-bit long, and enable processing up to eight data units per cycle at the same time. Normally, the *Xscale* datapath is used only for load/store operations, for loop counter management, and for other control tasks, while offloading computation to the *Wireless MMX* coprocessor. Systems like this generally provide high execution performance and have more structured on-chip memory architectures, even though the programming model turns out to be more complex and code portability properties are degraded. Moreover, a wide range of trade-off points in the energy-performance design space is usually made available (e.g., aggressive voltage and frequency scaling).

Another solution consists of using DSP-like coprocessors, thus providing the needed execution performance for most signal processing applications, but running the risk of not meeting the requirements of more computation-demanding multimedia applications. The architecture is in this case tailored to signal processing, and this makes it easier to optimize application execution and to meet its real-time constraints. This kind of architecture is well consolidated and its roadmap well defined. The support for software development is usually adequate, even though this is not able to counterbalance the increased complexity of the programming model, which resembles that of multiprocessor systems.

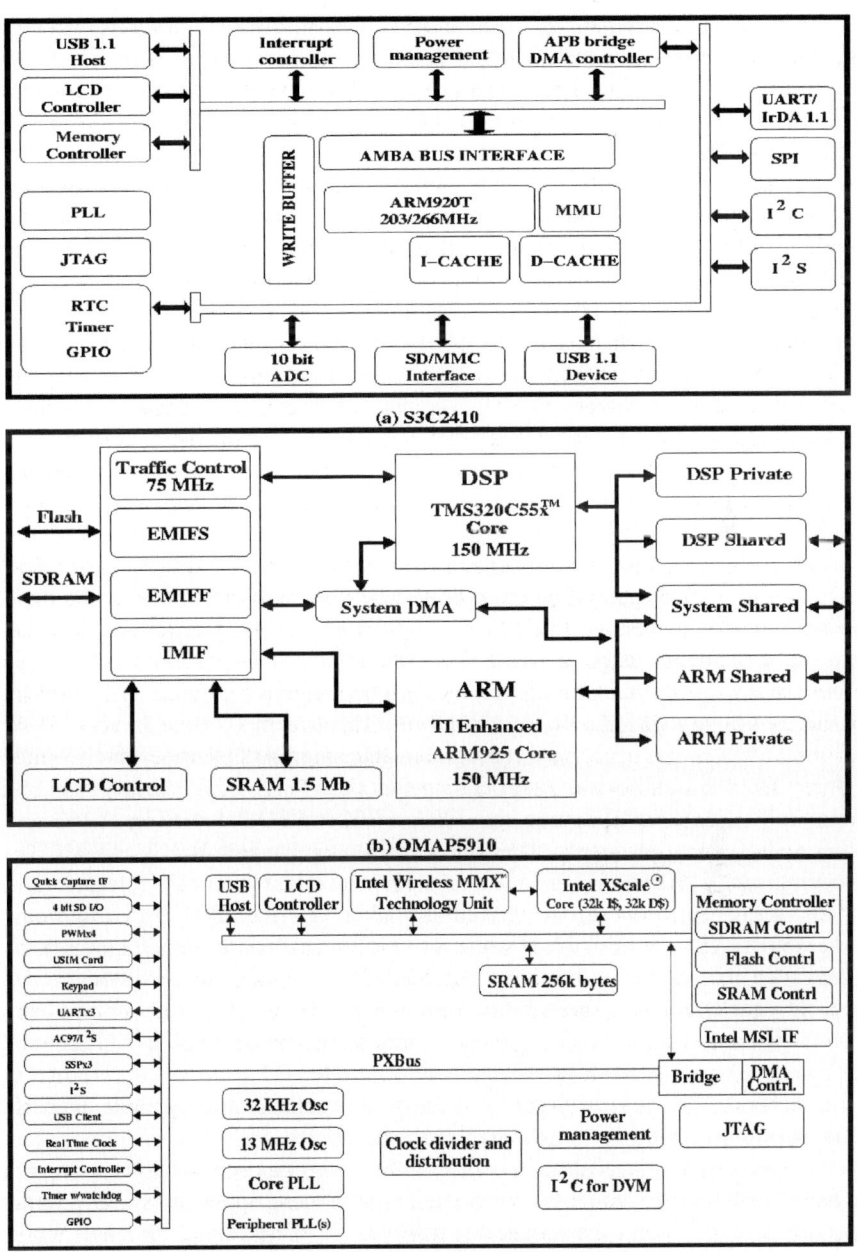

(a) S3C2410

(b) OMAP5910

(c) PXA27x

Fig. 1.1 Example of commercial processor cores used in mobile multimedia platforms

The integrated *OMAP5910* platform from Texas Instruments is an example of this approach [14]. It is a dual-core integrated circuit, with a general purpose *ARM925T* processor core working at 150 MHz (with a 32-bit multiplier and data-dependent throughput) and a *C*55 DSP working at the same clock speed and able to execute up to two 16-bit multiply-accumulate operations in a single clock cycle (Fig. 1.1b). The processor cores interact with each other by means of mailbox registers, the shared memory, or DMA transfers. The software development toolchain is very advanced, in particular in terms of APIs for the general purpose and the DSP processor cores.

This kind of solution usually provides good energy efficiency, moderate cost, and efficient memory usage, since the system is designed for signal processing from the ground up. In contrast, the dual-core architecture (and multi-core in general) makes programming an intricate task. The most critical aspect regards application partitioning, which might give rise to different trade-off points between performance and energy dissipation, and which cannot rely on mature design methodologies and CAD support yet.

An even more radical approach to performance and energy optimization for the execution of multimedia applications consists of using hardware accelerators. They usually require general purpose processors for the management of the multimedia content to be processed, for the control of system resources, and for interaction with the external world. The accelerators can be classified as *programmable* or *hardwired*. The former ones can be found, for instance, in the STMicroelectronics *Nomadik* multimedia platform [15]. It exploits the combination of an *ARM926EJ* processor core with programmable smart accelerators, which jointly manage the main audio and video coding functions (Fig. 1.2a). An audio accelerator is in charge of decoding the MP3, SBC, MPEG-2 Layer II, AAC, and other audio coding standards. The video accelerator handles H.263 and MPEG-4 coding, while H.264 will be made available in future Nomadik releases. The smart audio accelerator can be viewed as a C-programmable DSP core (a complete set of APIs is available), while the video accelerator consists of a programmable part and a customized part for hardwired functions. Although limited, the accelerator programmability aims at supporting coding formats and constantly evolving communication protocols without the need to modify the hardware architecture.

In general, this kind of solution involves customized programming models, toolchains tailored to the specific hardware accelerator, and poor software portability. Often, system level designers have to manage complex platforms with still immature software design tools. The combination of these accelerators with general purpose processor cores allows the execution of mobile terminal functions while digital signal processing is in progress.

Finally, it is possible to use completely hardwired accelerators without any programming support. This choice results in maximum performance and energy efficiency, but lacks flexibility. Sometimes it also leads to reduced clock frequencies due to hardware implementation trade-offs.

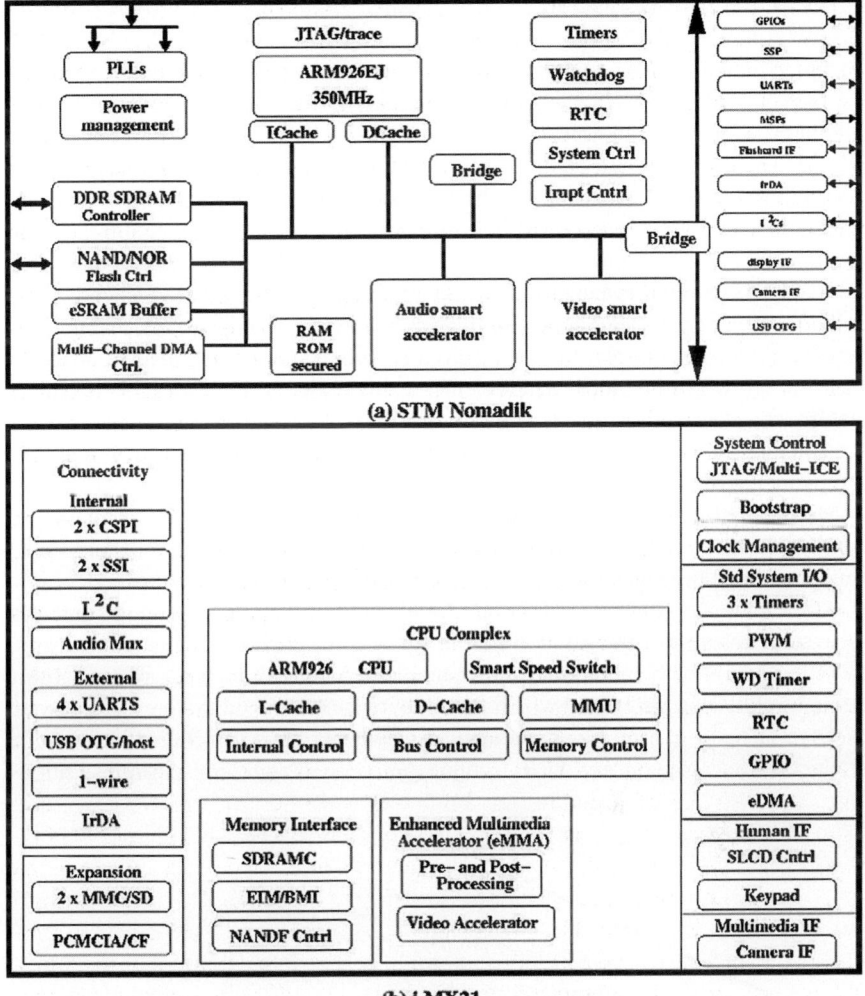

Fig. 1.2 Other commercial processor cores for mobile multimedia platforms

This is the solution employed for instance by the *i.MX21* processor from Motorola, which is the seventh generation of the *DragonBall* family of microprocessors for wireless portable devices [16]. It is based on the *ARM926EJ-S* core and on the *eMMA* (enhanced multimedia accelerator), as illustrated in Fig. 1.2b. It implements four fundamental functions in hardware: MPEG-4 and H.263 coding, preprocessing (colorspace conversion, formatting, image resizing), and postprocessing (deblock, dering, resizing, colorspace conversion). In order not to overload the CPU, the accelerator accepts input video streams by means of an internal, private, and dedicated data interface.

1.3 The Memory System

Designers of multimedia-enabled cell phones have to tackle the challenge of designing high density and high access speed memory systems. In the past, cell phones with basic wireless communication functionality used from 4 to 8 MB of low-power SRAM. The code required about 16 MB of NOR Flash memory or E^2PROM. With mobile terminal size reduction and the increase of offered services, the *multi-chip packaging* (MCP) technology was developed, which was able to mount in a unique package a 4 MB SRAM die and an 8 to 16 MB NOR Flash integrated circuit. This led to the reduction of space requirements for cell phone memories [17].

The advent of advanced applications such as Internet web browsers, text or multimedia messaging, games, and acquisition or processing of digital images led to a further growth of memory requirements, as illustrated in Fig. 1.3 [18]. A typical mobile terminal for cell phones makes use of 8 to 16 MB of low-power SRAM for data backup; 32 to 128 MB of Pseudo-RAM (PSRAM) for the system working area; and 64 to 128 MB of NOR Flash for bootstrap code.

Moreover, from 128 to 256 MB or more of additional memory (typically, a NAND Flash) are required for application software and for storage of large data structures (e.g., high resolution pictures, long audio streams).

At the same time, the size of mobile terminals is scaling down, and the memory subsystem has almost the same available space as in previous device generations, if not less. As already mentioned above, technology provides a workaround for this through multi-chip packages, which mount entire memory subsystems in a unique package of reduced size. From a technical viewpoint, stacks of 9 dies in the same product are feasible (*stacked MCP* technology), even though semi custom solutions

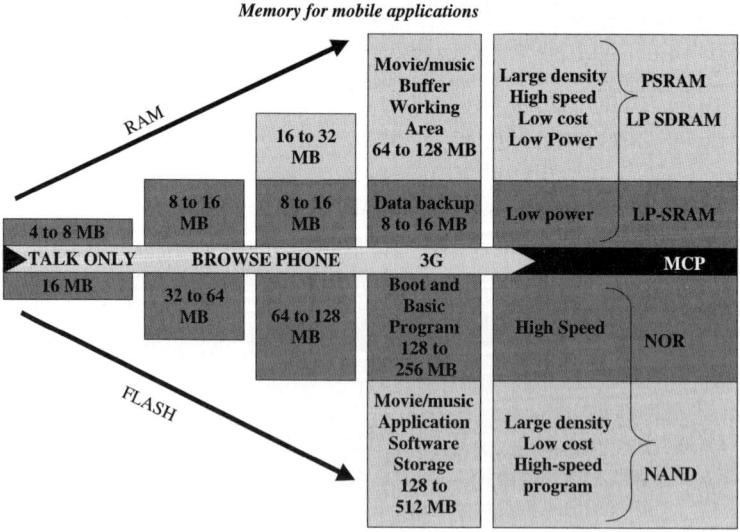

Fig. 1.3 Trends of Flash and RAM memory usage for mobile applications

with up to 5 or 6 ICs are normally used. It is worth pointing out that MCP uses a standard packaging technology such as *gold bonding wire*, which does not require additional investments for large-scale production.

From an architecture standpoint, there are two kinds of memory systems that are widely used across state-of-the-art mobile hardware platforms [19]. The first one consists of a high density and high performance variant of the traditional NOR+SRAM Flash memory architecture. The NOR Flash serves for code storage, and the SRAM for a working area. Alternatively, a solution employing low-power NAND Flash + SDRAM memory system is feasible, which achieves significant performance improvements. Code and data are stored in the NAND Flash, but are then stored in the SDRAM (working memory) in the preprocessing phase by means of a so-called *shadowing* architecture. This technique leads to performance speed-ups, which are counterbalanced though by the need to radically modify the embedded code with respect to traditional approaches, unless this functionality is directly supported by new generation chipsets. It is worth observing that the two alternative architectures have the same scalability properties, thus enabling designers to derive low-end, mid-range, or high-end terminals from the same reference multimedia platform without major changes in the memory architecture. In many cases only the memory cut or the stacked MCP configuration have to be changed. Let us now go into the details of the components building up the two memory architectures.

1.3.1 NOR Flash

NOR Flash memories have been long used as nonvolatile memory devices in common mobile terminals. Their access time is quite small (65 ns is a typical value), thus facilitating the *execute-in-place* (XIP) functionality. That is, the processing engine of the mobile terminal can directly fetch code from the Flash memory, instead of fetching it from a DRAM in which instructions have been previously stored from Flash memory.

The density of NOR Flashes was however not able to keep up with the increasing memory requirements of mobile phones. In fact, the relatively large cell size (about 2.5 times that of a NAND Flash) has caused a higher price for a NOR Flash with respect to a NAND Flash of the same capacity. Other disadvantages consist of the high programming time of large data blocks and of the much larger erase time than NAND Flash (e.g., 700 ms per block versus 2 ms). In future mobile phones, NOR Flashes are likely to be used as on-chip memory cores and in all those cases wherein code size is small (low-end platforms).

1.3.2 NAND Flash

NAND Flashes exhibit a larger density and generally achieve a higher capacity given the same area. Moreover, they have a cost per megabyte which is usually lower (from one third to one sixth that of NOR Flashes). Since however the random access in a NAND Flash is rather expensive, this kind of device is not suitable for XIP functionality and is usually employed for memories at higher layers of the hierarchy (similarly to hard disks in desktop PCs), and in combination with shadowing architectures. This is because the access methodology to the NAND Flash memory is serial (on a block basis) and not fully random.

Therefore, a typical access time for NAND Flashes approaches 50 ns for a serial access cycle, which increases to 25 µs for an initial random access cycle. These values have to be compared with the 70 ns required for a random access cycle in NOR Flash devices. When the data to be read is very large (e.g., an image acquired through a digital camcorder), the difference between NOR and NAND Flash performance is almost negligible, while it becomes significant for a typical "*program and erase*" sequence. With blocks of 64 KB, the NAND Flash clearly outperforms the NOR one: 33.6 ms versus 1.23 secs per block. More generally, NOR Flashes can be effectively used to frequently access small amounts of data, while NAND ones incur a high initial access time, but then enable fast access to long data sequences.

Many mobile terminals also provide removable memory cards to increase the total memory capacity. They are frequently based on NAND Flash memories, and are optimized for wide band access. Memory cuts can be as large as 2 GB, with transfer rates of 7.7 MB/s for reads and 6 MB/s for writes and a worst case average power consumption of 200 mW. In general, the memory capacity is not a delaying factor for the introduction of multimedia functionality in next-generation mobile terminals. The optimization targets are rather the access speed, the interface to the system, the cost per bit, and power consumption, as well as the testing methodologies during manufacturing.

NAND Flashes typically require that the chip-enable signal be activated during the entire read cycle. However, this prevents the processor from communicating with other devices connected to the same bus. For this reason, NAND Flash devices for mobile phones are equipped with a "Chip Enable Don't Care" mechanism, which allows the chip-enable signal to be deactivated during the read period. The processor can thus access other devices or I/O units in parallel.

1.3.3 Data Memory

The need to process and store multimedia content is leading to increasing SRAM memory cuts for use in mobile phones. SRAM cells exhibit very low power consumption due to the static bit storage mechanism, which avoids the need for refresh operations. A SRAM cell can be viewed as a flip-flop, composed of 4 to 6 transistors per cell. This relatively large cell size implies a high cost per bit. Pseudo-SRAM (PSRAM) devices have been designed to overcome this problem. They consist of DRAM memory cells made up of one transistor and one capacity for each cell. PSRAMs feature a lower cost per bit and usually integrate circuits for automatic refresh control. PSRAMs are available with similar access times to SRAMs (typically, 65 ns or less for random access) and support advanced mechanisms for page and burst mode.

A valid alternative to PSRAMs is represented by SDRAM (synchronous DRAM) devices, which are faster on average: their speed can be pushed to 83 and 133 MHz, to be compared with 65 ns for the asynchronous read access to PSRAM. Low-power SDRAMs (denoted by LP SDRAM) are available with larger capacities (up to 512 MB, versus 128 MB of PSRAMs).

Many processor cores integrate controllers for SDRAM memories, with operating frequencies of 133 MHz or more. However, an operating frequency larger than 100 MHz can hardly be supported by mobile hardware platforms, given their tight power budgets, although the high performance achievable by burst PSRAMs or LP SDRAMs will need to be fully exploited in next-generation high-end platforms. These latter will integrate multiple memory devices in multichip packages, and each device will have to meet different requirements (storage of executable code, program dedicated memory regions, communication memory, file system). At the time of writing, traditional NOR ι SRAM architectures are being replaced with the more efficient NOR+PSRAM+NAND architectures and even with the new NAND+LP SDRAM ones.

1.3.4 Integration Issues

Memory cores are the most frequently integrated components in SoC architectures, and a trend toward memory-dominated SoCs is underway. Presently, the impact of memory on SoC area ranges from 30 to 50%, but roadmaps envision that in 2011 SoCs will be memory dominated (about 90%), with a remaining 10% of the area consumed by programmable or reconfigurable logic. See Fig. 1.4 for details [20]. The integration of memories in SoCs (*embedded memory*) achieves clear advantages: power reduction and memory access speed-up, optimized multi-bank architectures for pipelined accesses and selective bank activation in low power mode, and optimized cuts based on application requirements. Memory integration cuts down on packaging cost, board area, electromagnetic interferences, chip area associated with I/O pads, and form factor of mobile terminals (which become smaller and more portable).

However, in some cases embedded memories may not be the best solution; in fact, standalone Flash or DRAM memories are mature and largely available products, and this leaves many degrees of freedom to system designers for price negotiation. As a consequence, in some products it may be more convenient to adopt MCP stacking technologies, piling up processor and memory cores in a single package. The memory cut can thus be specified in the latest design stages, when memory requirements posed by the software are known, and only the strictly needed memory size can be provided within the multichip package [21].

The integration of Flash memory in the same die with CMOS logic requires the harmonized combination of two very different manufacturing processes in a unique

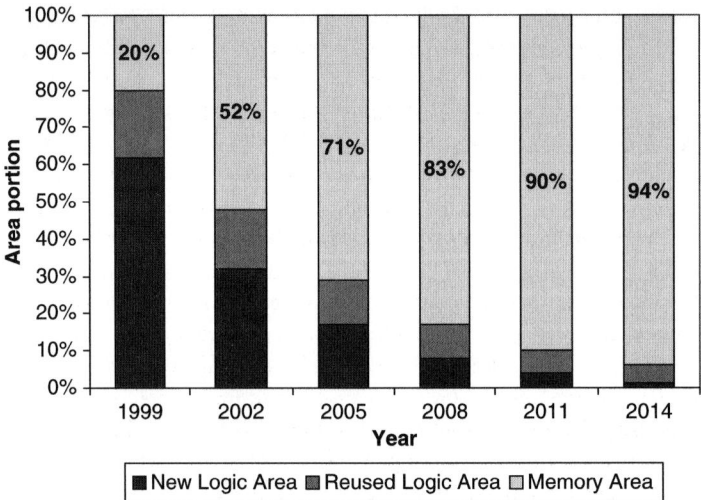

Fig. 1.4 Trend of the ratio between the amount of integrated logic and memory in Systems-on-Chip

procedure. This merging of technologies has resulted in several innovations in the traditional manufacturing process of Flash memories, which are described hereafter.

Trench isolation. This kind of isolation is required for ultra high integration density SoCs and for the layout of very small SRAM cells. Trench isolation is not commonly found in Flash manufacturing processes. For instance, Intel introduced it beginning with the 0.25 um process for Flash memories.

Differentiated gate oxides. These are required to support both the high-voltage program and erase operations of Flash cells and the reduced-voltage integrated CMOS logic operation. The 0.18um Intel Flash process supports differentiated gate oxides.

Low thermal budget. In order to obtain high performance transistors, process steps should use moderate temperatures. While Flash cells are usually integrated before CMOS logic, the opposite holds for DRAMs, and their high temperature process steps might limit final performance of logic functions.

Metal Layers. Multiple metal layers are required for highly integrated digital circuits. The increase of their number has not been a hindrance to the integration of different technologies, since their deposition is fully compatible with Flash and CMOS logic process steps.

Salicided gate. This kind of gate for PMOS and NMOS transistors is required to obtain low threshold voltages and reduced channel lengths, i.e., high performance logic functions. The integration of these gates with memories turns out to be difficult, and Intel has introduced salicided gates in the 0.25um Flash technology.

From a technology viewpoint, there are two possible approaches to the integration of memory and logic. If the memory cuts required by the system are significant and comparable to those of standalone memories, then it is convenient to adopt the memory manufacturing technology as the baseline process and to augment it with the variants needed for logic integration. This way, the reliability of Flash cells is preserved at the cost of a performance slow-down of CMOS logic with respect to *ad-hoc design* (up to one generation). In contrast, computation-intensive systems where required memory cuts are limited adopt the CMOS logic manufacturing process as the baseline process, to which the process steps needed to implement Flash cells are added. This approach results in reliability degradation and in unoptimized area occupancy for the Flash memory, which translates into a low integration density.

As far as DRAMs are concerned, the manufacturing process features substantial differences with respect to that for CMOS logic. This latter typically requires 1 or 2 layers of polycrystalline silicon and 6 or 7 metal layers, while DRAMs require many levels of poly-silicon and just 2 metal layers. Hybrid technologies therefore incur lower circuit density due to wire routing constraints. Moreover, DRAMs need 4 additional process steps in hybrid technologies, which are applied also to CMOS logic and therefore increase the number of masks and the wafer cost. In Table 1.1, the overhead incurred by hybrid technologies in terms of additional process steps is reported for a different mix of baseline manufacturing processes.

1.3.5 System Level Design Issues

System level design issues associated with the integration of memories concern partitioning, i.e., the definition of the multibank structure. This allows the dynamic management of power consumption, the reduction of wordline and bitline lengths, and the pipelining of accesses to different banks. In contrast, this architecture requires complex selection and control logic and larger area occupancy.

Moreover, since CPU performance increases more rapidly than DRAM performance, architecture-level design techniques are required to exploit the principle of locality. The most frequent solution consists of defining a memory hierarchy with increasing size and decreasing performance as we move across memory layers.

Computation data, which is likely to be accessed more frequently, is moved to fast access memories such as caches. The access penalty to higher level storage

Table 1.1 Additional process steps for hybrid technologies (2003 update)

Added process	Logic	SRAM	Flash	DRAM	CMOS RF	FPGA	MEMS
Logic	0						
SRAM	1–2	0					
Flash	4	3–4	0				
DRAM	4–5	3–4	7–9	0			
CMOS RF	3–5	5–9	6–9	6–10	0		
FPGA	2	2–4	4–6	3–7	5–7	0	
MEMS	1–10	3–12	6–14	6–15	5–15	4–12	0

devices in the hierarchy (performance- and energy-wise) is reflected in application performance depending on the amount of cache misses, which in turn is strongly dependent on data locality and on the structure and size of caches. An important cache parameter is *associativity*. The mapping of data to the cache is *fully associative* if a memory block can be found in any cache location. In a *set associative* mapping, a memory block can be found in any cache location within a given set determined by the address. Finally, a *direct mapping* can be implemented, wherein the cache location is univocally determined based on the address. Normally, miss rate reduction techniques consist of making data blocks larger, increasing cache associativity and compiler-driven code optimizations.

The most advanced microprocessors also support direct connection to *tightly coupled (or scratchpad)* memories. The direct connection between processor and scratchpad memory allows very low access cost for the processor. Interestingly, these memory cores are software-controlled, in that data storage is directly managed by the application and not by automatic replacement algorithms implemented in hardware, as for caches. The *ARM9E-S* family of microprocessors provides two interfaces for tightly coupled memories: one for data and the other for instructions.

While cache performance mainly comes from the precision with which the replacement algorithm predicts future memory access patterns, scratchpad memories rely on the full knowledge that applications have of these patterns. A higher predictability of execution times follows from this, which makes scratchpad memories suitable for real-time embedded systems. Their main drawback lies instead in the need to make explicit data copies from an on-chip memory, and in their limited size, which might make several processing data swapping operations necessary when the data footprint is large.

Caches and scratchpad memories usually coexist in real-time systems: data accessed throughout the entire application execution or with high temporal locality are stored in scratchpad, while data structures exceeding scratchpad size and featuring temporal or spatial locality typically reside in cache.

Scratchpad memories represent the first step toward the implementation of distributed memory systems, for which message passing represents an intuitive programming model. In this case, systems equipped with scratchpad memories provide an additional degree of freedom, namely the implementation of message exchange queues directly in scratchpad.

1.4 The Display

The display is one the most expensive components in a cell phone, even though its impact on sales justifies its cost. Display technology is evolving fast, as pointed out in Fig. 1.5. Quarter-VGA displays (QVGA, about 77000 pixels of resolution) introduced in 2003 became mainstream in 2005/2006, while they were already a standard in Japan and Korea [22].

Fig. 1.5 Trend of display resolution in mid-tier mobile phones (Source: EMP)

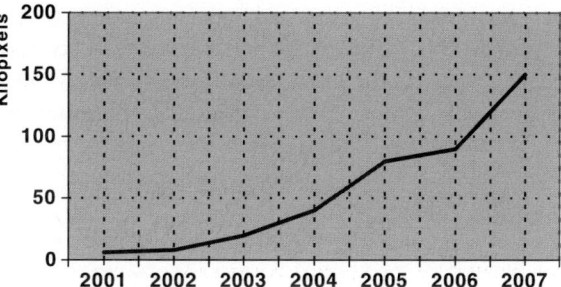

Pixel density in cell phone displays is higher with respect to laptop computer or desktop PC displays. Laptops have from 100 to 135 pixels per inch (PPI), while high-end cell phones range from 150 to 200 PPI. Some prototypes achieve 200–400 PPI and will arrive in the market in the form of 2/2.5 inch VGA displays (0.3 megapixels). In spite of the excellent quality of these displays and their support for advanced graphics applications, many users will still be able to capture the pixel-level granularity of images. In fact, the limit resolution of the human eye is approximately 700 PPI at a viewing distance of 25 cm. Note that above-standard printers easily exceed this resolution, making reading of printed texts preferable with respect to their display visualization.

There are ample margins for optimizing the energy efficiency of displays [23]. LCD technology is based on light transmission modulation, and resembles a light valve. The quantity of white light, which goes through a liquid crystal, is varied, thus obtaining a grey scale. The light polarization follows the orientation of particles in the liquid, which has optical properties similar to those of a solid crystal. Particle orientation is regulated by means of an electric field. Light goes through two polarized filters, having an angle shift of 90 degrees, depending on the electric field applied to the liquid crystal in each cell. In case of a color LCD, three colored filters (red, green, blue) are associated with each image element. In active matrix TFT displays, the single image elements are driven by means of transistors, thus increasing the switching speed of the device. The TFT panel is placed on top of a backlight panel. All pixels on the TFT LCD panel are backlighted. The polarizer and color filters shutter and modulate the backlight, thus making a pixel more or less luminous. This method of producing an image is highly inefficient, since more than 90% of the backlight intensity is usually lost.

A different approach is taken by the organic light emitting diodes (OLEDs). They consist of pixels or electroluminescent elements made up of organic material, which directly emit light when they are crossed by a current. This leads to a power reduction (from Ws to mWs), a better brightness and contrast (from 25:1 to 100:1), a better view angle (from 100 to 160 degrees), and better response times with respect to present TFT displays. However, problems associated with different life times of emitters, residual charge in the parasitic capacitances of display elements, low manufacturing yields and high cost, make OLED a technology for niche products (e.g., small displays for home appliances, secondary displays for high-end cell phones, displays for car radios). This technology is however very promising for its high energy efficiency, and a lot of companies are involved in the development and production stages (Pioneer since 1997, TDK since 2001, Samsung-NEC since 2002, RiT display since 2003, and Sanyo-Kodak since 2003).

1.5 The Transceiver

In January 1998, the European Institute for Telecommunication Standards (ETSI) selected the wideband CDMA technology (W-CDMA) as the multiple access technique for third-generation mobile phones. We will show an overview of the main aspects and of the main implementation issues of a CDMA transceiver, which is an essential component of each mobile multimedia platform for cell phones. The structure of a generic W-CDMA transceiver consists of a radio-frequency section (RF), an intermediate frequency section (IF), and a baseband section. The RF and IF sections are typically implemented with analog technology, while the baseband section usually features digital implementation. However, the analog-to-digital interface tends to move more and more toward the antenna.

1.5.1 W-CDMA

W-CDMA communication systems rely on transmission channel sharing among different communication flows not by means of time or frequency division multiplexing, but rather by means of proper codes, which have a one-to-one correspondence with the communication flows. The underlying assumption is that these codes are known from the communication actors in some way. Moreover, the band of the transmitted signal is enlarged with respect to the band that would derive from the direct transmission of the modulated signal in the channel. At the physical layer, this occurs through the *spreading* operation: the information sequence is multiplied by a spreading sequence consisting of a periodic sequence of binary symbols (called *chip*), which can have a $+1$ or a -1 value. The chip transmission frequency is 3.84 Mcps in UMTS. A *spreading factor* then divides the chip frequency to obtain the actual transmission frequency, and therefore provides a degree of freedom to vary the transmission bit rate.

The spreading operation serves to distinguish different transmissions generated by the same transmitter. A particular spreading sequence, different from that of other users, is associated with each communication flow. The same thing happens when a mobile terminal has to transmit at the same time more traffic flows to the base station: each flow uses a different spreading sequence. The selected codes are orthogonal with each other: given n orthogonal codes, it is possible to transmit n signals from the same transmitter at the same time, without any mutual interference. The receiver performs the inverse *despreading* operation, which also results in band shrinking. In practice, the received wideband sequence is multiplied by a spreading sequence (known from the receiver) and the resulting signal is integrated over a symbol time through an adapted filter. This reconstructs the original signal.

In the context of mobile phones, transmissions from multiple transmitters at the same time have to be handled as well. The ideal solution would be to have all traffic flows from each source transmitted by means of orthogonal codes, but this is not feasible due to synchronization issues and to the limited number of codes. Therefore, a scrambling operation is performed by each terminal, which multiplies the output of the spreading process by a new scrambling sequence (at the same frequency), which is much longer than the spreading sequence. The receiver will just have to multiply the received signal by the same scrambling sequence in order to reconstruct the signal before despreading is applied.

Finally, let us consider the effect of reflected waves (*multipaths*) in a W-CDMA system. First of all, several contributions of the transmitted signal will arrive at the receiver, each of them delayed by a multiple of the chip time. For a first approximation, the despreading process allows the detection of one of the delayed contributions of the signal, and the interference of the other ones can be neglected. In TDMA systems, on the other hand, the presence of multiple paths requires the use of complex equalization systems. Moreover, due to the well-known *fading* phenomenon, the amplitude of each contribution has to be considered as random, and the power of each received signal may incur significant fluctuations around a given average value. This would imply a severe degradation of receiver performance. As a workaround, a particular kind of receiver known as *RAKE* is used. It is made up of a certain number of parallel correlation-based receivers (called RAKE fingers), each one tuned to a particular contribution of the signal. The receiver then combines the output signals of each adapted filter before taking a decision on the bit. This way, the amount of received signal energy is maximized, thus increasing the signal to noise ratio. Moreover, the combination process alters the statistics of the received power by limiting power fluctuations around the average value.

Finally, the development of proper real-time power control systems is what made the adoption of W-CDMA feasible. In fact, in the absence of such control, the signal coming from a nearby point would cover those coming from far away points even after the despreading, making their reception impossible. This explains the use of a higher power by base stations for the furthest users, so that all of them can experience a similar signal-to-interference ratio. This objective is achieved by means of a power control made on each downlink.

In W-CDMA it is necessary, when possible, to reuse the same frequencies in adjacent cells in order to avoid temporary disconnections during handoffs, which are not tolerable, in particular for high-speed data transmissions.

1.5.2 W-CDMA Modem Architecture

A generic architecture for a W-CDMA transmitter and receiver is illustrated in
Fig. 1.6 [24]. The baseband transmitter is mainly responsible for coding information
to be transmitted and for signal modulation and spreading. Algorithms for error
control can be classified in schemes based on *data blocks* or *trees*. Linear codes
(Hamming, Reed-Muller), cyclic codes (Golay, BCH, Reed-Solomon) and burst
error detecting codes (Fire, interlaced codes) are examples of the former class of
techniques, while convolutional codes are the most famous tree-based schemes. Re-
cursive convolutional codes or Turbo codes are also frequently used.

In third-generation mobile terminals, encoders must be able to work at different
speeds and, to their limits, support different coding algorithms. The frequency of
data provided by the system is made compatible with the data rate of the radio interface by means of the symbol repetition technique. Control information or header
bits are coded and inserted into the stream, together with power control information
and the necessary number of reference symbols.

After coding, data modulation follows, such as BPSK or QPSK, which provides
a good spectral efficiency. Spreading is then applied to modulated components, to
the one in phase or to both the ones in quadrature, based on the kind of modulation.
This operation reduces to a binary multiplication performed in hardware. A filter
precedes the D/A converter to obtain the required transmission band. In the analog RF section of the receiver, automatic gain control (AGC) is implemented and

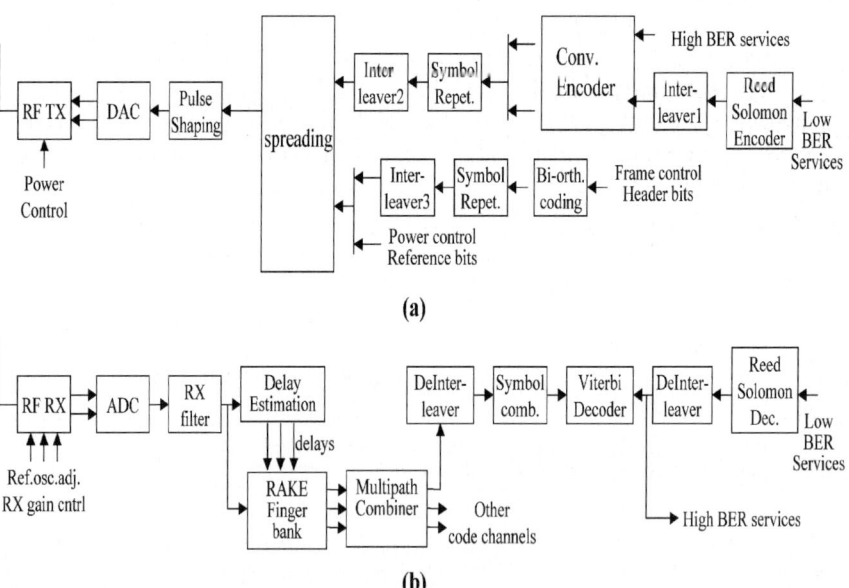

Fig. 1.6 (**a**) Transmitter and (**b**) receiver sections of a W-CDMA modem; the RF and baseband
sections are illustrated

some feedback signals from the analog section provide an error signal to tune the reference oscillator for the purpose of clock synchronization. After analog-to-digital conversion and low-pass filtering, the wideband stream is then fed to the despreading unit, where a bank of correlators changes the input into a narrowband signal. Then, all multipaths from the finger outputs of the RAKE receiver are combined to form a composite signal that is expected to have substantially better characteristics for the purpose of demodulation than just a single path. In order to tune each finger on a different multipath, it is necessary to provide an estimation of channel delay profile. The optimal combination of the received components is called *maximal ratio combining*, and consists of summing the squares of the signal contributions. The receiver finishes by applying deinterleaving and decoding to the symbol sequences for which the user owns the spreading code.

1.5.3 Cost Metrics

Because of the fast evolution and miniaturization of units for digital signal processing such as ASICs and DSPs, the radio frequency analog section is typically the most power consuming section in W-CDMA terminals. This trend is partially offset by the increasing complexity of the digital processing architecture and by the shrinking of the analog section. Power consumed by this latter depends on the data rate, on the activity factor, on actual irradiated power, and on linearity requirements. Power amplifiers are certainly the most power hungry components and are becoming critical to meet the increasing linearity requirements on wider bands. In contrast, the size of radio cells is decreasing (pico/micro-cells); therefore transmission power can be reduced and overall power can be cut down accordingly. Other power hungry components include frequency synthesizers, which burn a power directly related to the activity factor and are therefore larger for continuously transmitting systems like CDMA ones. Moreover, two frequency synthesizers are generally required, since transmitting and receiving functions are carried out simultaneously, contrarily to TDMA terminals, which need a single synthesizer.

The analog and digital processing functions usually do not play a major part in determining terminal size. Dominant contributors are instead batteries and the user interface. However, some radio frequency components (such as the duplex filter and the antenna) take a significant space. On the contrary, the area of the digital section tends to be progressively reduced and collapsed into a unique integrated circuit, thus providing maximum area savings. Isolation requirements and integration issues of hybrid technologies prevent the integration of the radio frequency front end in the same silicon die.

Third-generation W-CDMA terminals provide a set of very heterogeneous services, ranging from low bit rate telephony services to high bit rate and high quality data services. This poses a challenge to the architecture, which should be able to adapt to the specific service being offered, thus determining a high efficiency in available resource utilization.

In a W-CDMA system, high transmission data rates can be achieved by decreasing the spreading factor or by introducing more parallel channels. This latter option requires high peak-to-average power ratios and thus conflicts with the use of power amplifiers in transmitters. It can be therefore more convenient to use parallel channels in the downlink to obtain a good modularity, and a variable spreading factor in the uplink to avoid parallel transmissions in the mobile terminal. The baseband section features a complexity which is almost independent of service data rates, which are only associated with the integration time of correlators in data fingers. However, the use of parallel channels poses additional requirements on components such as the pulse-shaping filter, RAKE fingers (which should be replicated) and code generators. This cost is mitigated by the fact that other components need not be replicated (receiver front end, multipath delay estimator, etc.).

Components operating on narrowband signals are not as scalable as those in the wideband section. The use of multiple channels unavoidably requires a higher allocation of hardware resources, mainly for coding, decoding, and interleaving. The needed scalability can be provided in two ways. On the one hand, multiple processing engines can be activated in parallel, exploiting area availability in nanoscale technologies. On the other hand, a higher throughput of processing engines can be achieved by increasing their clock frequencies. Obviously, the trade-off between computational horsepower and power consumption should be carefully assessed. For instance, it is not desirable to have high operating frequencies in battery-operated mobile terminals.

1.5.4 Software Radio

The most flexible implementation of a mobile terminal would consist of its full software configurability. This approach would allow the access to different wireless systems through a unique hardware platform by internally uploading the new terminal configuration or by downloading it from the network. There are a few main challenges in designing this kind of terminal. The radio frequency section should be able to transmit and receive signals with different modulation techniques and with different frequency bands. This requires a linear power amplifier on a wide range of frequencies with acceptable efficiency. Similarly, antennas should feature low losses and uniform gains across the same range of frequencies.

After the down-conversion to IF or to baseband, it would be necessary to select the signal in the desired band before A/D conversion is performed, in order to bring back the required dynamic range to the converter within acceptable limits. A solution would consist of using a bank of filters and of selecting a specific band based on the terminal operating mode. The baseband section of a software radio would be made up of programmable hardware and powerful DSP processors. In the former case, programming would be carried out through FPGA technology and consolidated hardware description languages such as VHDL and Verilog. Based on the operating mode, different features should be supported in baseband. The major problem concerns channel equalization and FEC (de)coding. In fact, both the burst-oriented equalization in TDMA and the multipath combination in DS-CDMA should be supported efficiently. In general, a software implementation of channel coding/decoding is effective when the actual bit rate is low, while hardware accelerators are needed to deal with high bit rate services and operating modes.

Finally, an important issue raised by software radio concerns the flexible utilization of available computation resources. Given the difficulty in increasing the clock frequency of a mobile terminal, additional computation resources could be obtained by instantiating and activating them on reconfigurable logic.

1.5.5 Multi-Mode Terminals

A big challenge in designing modems for mobile terminals consists of the need to support different radio technologies. The main third-generation cellular systems include GSM, IS-136, IS-95, and PDC. Third-generation W-CDMA systems are implemented together with some of these systems, for instance GSM/W-CDMA, PDC/W-CDMA, and IS-95/CDMA2000.

Each kind of system poses different requirements on the transceiver components. As regards the RF, dual-mode terminals have to operate at different frequencies, thus necessitating different tuning bands of the RF and IF filters, requirements on frequency synthesizers, and A/D and D/A converters. The baseband reception algorithms are also different. A system based on TDMA requires an equalizer, while a CDMA-based one needs a RAKE receiver.

A further differentiation arises for *slotted* versus *continuous* systems. The latter are more suitable for ASIC implementations, since most of the functionality is carried out continuously. In this case, a software implementation would cause an overly large control overhead for each received symbol, because it should be processed independently through the generation of an interrupt for a DSP. This overhead might be reduced by prebuffering symbols. For these reasons, software implementation best matches the features of slotted systems: symbol detection starts only when the entire burst has been received. In the RF section, power amplifiers undergo discontinuous transmissions with high peak power in slotted systems, while operation is constant but with reduced peak power in continuous systems.

1.6 The Bus

The communication architecture is becoming the key bottleneck to achieving the performance envisioned for highly integrated SoCs (see for instance the processors in Figs. 1.1 and 1.2). The main challenge lies in the scalability of communication fabrics, since the number of integrated cores with communication requirements on the bus keeps increasing, pushed by technology scaling and by the need for more computation resources.

Typically, SoCs make use of shared buses to accommodate communications among cores, based on a serialization mechanism of bus access requests (arbitration), which solves access conflicts to the shared medium. This solution exhibits a number of inefficiencies when put to work in highly integrated systems. First, it leverages a broadcast communication technique which charges/discharges the input capacitances of all modules connected to the bus at each bus transaction. This leads to a useless waste of energy and to a performance slowdown in the presence of a large number of integrated modules. However, the architecture of shared buses is simple and results in limited area occupancy, thus motivating the widespread adoption of this solution for SoC designs with moderate complexity. The most recent SoCs adopt communication architectures with more advanced features. On the one hand, ongoing efforts are aimed at increasing the efficiency of communication protocols, so that a large percentage of bus busy time can be spent for actual data transfers, thus reducing the overhead of other protocol-related mechanisms (e.g., arbitration cycles, bus handover overhead). On the other hand, a trend is underway to increase the bandwidth made available by the communication fabric by implementing parallel topologies, such as crossbars or bridged buses.

Some preliminary explorations indicate that, even with more efficient protocols and more parallel topologies, highly integrated SoCs will incur bus saturation effects [9]. This calls for disruptive interconnect technologies addressing the scalability issue from the early design stages, and not as an afterthought. On-chip interconnection networks (Network-on-Chip, NoC) are generally believed to be the long-term solution to the problem of on-chip communication, and they rely extensively on scalability and modularity in interconnect design [25].

Communication architectures are produced by leading semiconductor industries (e.g., CoreConnect from IBM, STBus from STMicroelectronics), by processing core manufacturers (e.g., AMBA bus from ARM), or by companies specifically targeting interconnect design (e.g., SiliconBackplane from Sonics) [26]. In order to shed light on bus solutions for integrated systems, we will provide an overview of a successful architecture, which is integrated into many low- to medium-end industrial platforms: the AMBA family of on-chip interconnects.

1.6.1 AMBA Bus

The AMBA (Advanced Microcontroller Bus Architecture) standard was originally conceived to support communication between ARM cores, but it has rapidly become a de facto standard for industry-relevant high-performance embedded microcontrollers.

The AMBA specification defines a shared bus topology, but also envisions a low-complexity segmented architecture. In fact, two main and distinct buses are defined within the specification. The AMBA Advanced High-Performance Bus (AHB) acts as the high-performance system backbone for high clock frequency system modules. AHB supports the efficient connection of processors, on-chip memories, off-chip external memory interfaces, and high performance peripherals.

The AMBA Advanced Peripheral Bus (APB) allows the interconnection of the processor to lower power and less performing peripherals, hence provides simpler communication mechanisms and lower bandwidth. AMBA APB can be used in conjunction with the AHB system bus by means of a connecting *bridge*, as illustrated in Fig. 1.7a. The main features of AMBA AHB can be summarized as follows:

Multiple bus masters. A simple request-grant protocol and a flexible arbitration module with user-defined contention resolution algorithms allow sequential bus access for multiple communication initiators (*bus masters*).

Pipelining and burst transfers. A bus transaction consists of an addressing phase and a data phase. Both phases take place during consecutive clock cycles. Communication pipelining enables the overlapping of the addressing phase of one transaction with the data phase of the previous transaction served on the bus, and results in higher throughput. AMBA AHB also supports burst transfers: multiple data words can be transferred across the data bus with a single arbitration round.

Split transactions. When a slave (i.e., a communication target) cannot complete a transfer immediately or in a short time, the bus can be released by means of a split mechanism. The master is frozen and bus ownership will be requested by the slave on behalf of the master when the transfer can complete. The split mechanism ensures that other masters are prevented from accessing the bus for long periods of time in the presence of high latency slaves.

A generic transfer on the AMBA AHB bus consists of 4 cycles. During the first one, the master drives a request signal to the arbiter. In the best case, the arbiter responds with a grant during the second cycle. Then, the addressing cycle and the data transfer cycle follow. Four-, eight-, and sixteen-beat bursts are defined in the AMBA AHB protocol, as well as undefined-length bursts and single transfers. Bursts are an effective way to cut down on arbitration overhead.

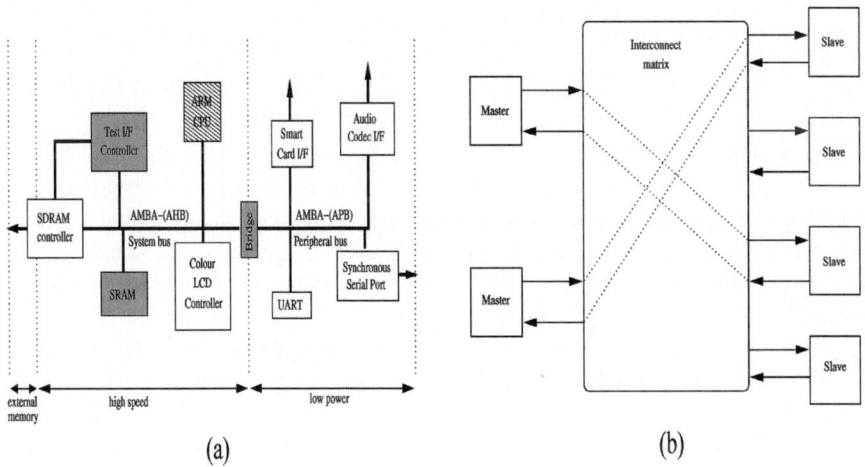

(a)   (b)

Fig. 1.7 (**a**) AMBA AHB, (**b**) Multi-Layer AHB

1.6.2 Multi-Layer AHB

Multi-Layer AHB is an interconnection scheme based on AMBA AHB protocol that enables parallel access paths between multiple masters and slaves in a system. In practice, the specified architecture goes in the direction of increasing the available bus bandwidth through an interconnect fabric supporting communication parallelism. This is achieved by using an interconnection matrix, which allows masters to access separate slave cores in parallel (Fig. 1.7b). A key advantage of Multi-Layer AHB is that standard AHB master and slave modules can be used without the need for modification. When several masters want to access the same slave at the same time, arbitration logic at the target output port of the interconnection matrix solves the conflict. One input port of the matrix can be connected to an individual AHB master interface (which can be simplified since no arbitration nor master-to-slave muxing is required) or to a full AHB layer. In spite of the higher degree of parallelism enabled by multilayer AHB-compliant system interconnects, the latter suffer from the same hardware scalability limitations of crossbars and might be infeasible for systems with more than 10 masters and 10 slaves.

1.6.3 AMBA AXI

AXI is the latest generation of AMBA interfaces. It is targeted to high-bandwidth and low-latency designs, enables high frequency operations without using complex bridges, is suitable for memory controllers with high initial latency, and is backward compatible with existing AHB and APB interfaces. AXI aims at a better exploitation of the bandwidth made available by the interconnect fabrics by means of protocol optimizations.

The main innovation introduced by AXI lies in the specification of a point-to-point connection. It decouples masters and slaves from the underlying interconnect by defining master interfaces and symmetric slave interfaces (Fig. 1.8a). AXI interface definition allows the description of interfaces between a master and the interconnect, a slave and the interconnect, and directly between a master and a slave. This approach ensures interconnect topology independence and simplifies the handshake logic of attached devices, which only need to manage a point-to-point link. While AXI interfaces enable a variety of different interconnect implementations, previous AHB protocols require an exact knowledge of implementation details. As an example, the AHB arbiter is instance-specific, in that it is connected to master interfaces by means of request and grant signals.

To provide higher parallelism, four different logical mono-directional channels are provided in AXI interfaces: an address channel, a read channel, a write channel, and a write response channel. Activity on different channels is mostly asynchronous (e.g., data for a write can be pushed to the write channel before or after the write address is issued to the address channel) and can be parallelised, allowing multiple outstanding read and write requests, with out-of-order completion. AXI slave

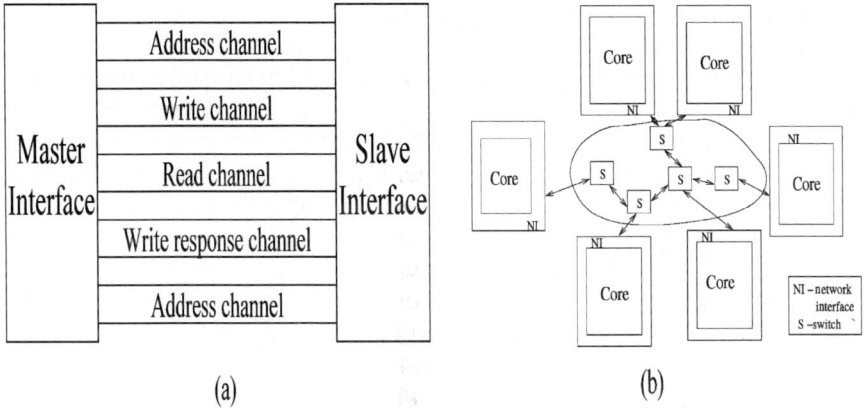

Fig. 1.8 (**a**) AMBA AXI, (**b**) Network-on-Chip

interfaces use the write response channel to notify the master interface of the success or failure of a write transaction

The mapping of AXI channels to instance-specific interconnect resources is decided by the interconnect designer. Most systems use one of three interconnect approaches: shared address and data lanes, shared address and multiple data lanes, or multilayer with multiple address and data lanes. In most systems, the address channel bandwidth is significantly less than the data channel bandwidth. Such systems can achieve a good balance between system performance and interconnect complexity by using a shared address lane with multiple data lanes to enable parallel data transfers.

Although we used AMBA as a case study for the trends in on-chip communication, the landscape of evolutionary interconnects with respect to shared buses is very diverse, and many other alternatives do exist.

1.6.4 Network-on-Chip

In spite of protocol and topology enhancements introduced by state-of-the-art interconnects, evolutionary approaches do not address in full the fundamental scalability limitations of any single-hop interconnect. In the long run, more radical solutions are needed.

Networks-on-Chip are generally viewed as a disruptive interconnect technology able to tackle the challenges of on-chip communication in highly integrated multiprocessor System-on-Chip platforms. NoCs are packet-switched multihop interconnection networks integrated onto a single chip. Guiding principles and communication mechanisms are borrowed from wide area networks and from off-chip interconnection networks used in large-scale multiprocessors. However, the on-chip setting poses unique challenges in terms of power consumption, communication performance, area occupancy, and fault tolerance. With respect to previous embodiments of the network concept, NoCs differ in their local proximity and their lower nondeterminism.

In NoCs, communicating cores access the network by means of point-to-point interfaces (e.g., AXI, OCP). Transaction information is used by network interface modules to perform two fundamental tasks: communication protocol conversion (from end-to-end to network protocol) and packetization. Usually, clock domain crossing also occurs in network interfaces to decouple core frequency from network speed. Packets are then forwarded to their destinations through a number of switching elements, implementing multihop communication.

NoC architectures enable a scalable and modular design style, since they are based on multiple replications of two basic building blocks: switches and network interfaces (Fig. 1.8b). The interconnection pattern of the switching elements denotes the network topology, which can be regular (e.g., mesh, torus) or irregular (e.g., customized domain-specific topologies). Topology should be selected in the early design stages and plays a major role in determining average packet latency and overall communication bandwidth.

Other key design choices for on-chip networks include routing algorithm and flow control technique selection. The objective of routing is to find a path from a source node to a destination node on a given topology. The goal is to reduce the number of hops for communications across the network and to balance the load of network channels. Flow control determines how network resources (such as channel bandwidth and buffer capacity) are allocated to packets traversing the network. Inefficient implementation of flow control has side effects in terms of waste of bandwidth and unproductive resource occupancy, thus leading to the utilisation of a small fraction of the ideal bandwidth and to a high and variable latency. All flow control techniques using buffering need a means of communicating the availability of buffers between switching elements. This buffer management logic informs the upstream switch when it must stop transmitting packets because all downstream switch buffers are full (backpressure mechanism).

An intensive worldwide research effort and the first industrial prototypes are pushing the fast evolution of NoC design principles and practice. The main challenges that need to be tackled to make NoC technology become mainstream include power and area optimizations, the development of novel design methodologies (both front-end and back-end), and the definition of suitable programming models for network-centric hardware platforms.

1.7 Conclusions

This chapter has provided an overview of the main building blocks in nomadic multimedia platforms, with emphasis on cell phone applications. Each system component (from processing engines to memory cores, from transceivers to system interconnects) has been described, pointing out the main design issues and emerging trends. We have observed a clear trend toward the introduction of advanced multimedia services in mobile platforms, which poses increasing computation power requirements with architecture-level implications for the entire system. Since nanoscale technologies enable the integration of an increasing number of cores onto the same silicon die, the computation scalability is being addressed by multicore chips. Functions are divided up into parallel tasks which are carried out by a number of processing engines operating in parallel. This way, core operating frequencies can be kept low, resulting in energy efficiency and computational power scalability. Multicore chips are being introduced not only in high performance microprocessors, but also in the embedded system-on-chip domain. This trend poses new challenges to the design of the memory architecture and of the system interconnect.

References

1. K. Tachikawa, Requirements and Strategies for Semiconductor Technologies for Mobile Communication Terminals, *Electron Devices Meeting*, pp. 1.2.1–1.2.6, 2003.
2. R. Want, T. Pering, G. Borriello, K.I. Farkas, Disappearing Hardware [ubiquitous computing], *IEEE Pervasive Computing*, Vol. I, no. 1, pp. 36–47, 2002.
3. F. Boekhorst, Ambient Intelligence, the Next Generation for Consumer Electronics: How will it Affect Silicon? *IEEE Int. Solid-State Circuits Conference*, pp. 28–31, 2002.
4. J.Ostermann et al., Video Coding with H.264/AVC: Tools, Performance and Complexity, *IEEE Circuits and Systems Magazine*, Vol. 4, no. 1, pp. 7–28, 2004.
5. *Qualcomm CDMA technologies*, Qualcomm WCDMA (UMTS) 3G Chipset Solutions, *online: www.cdmatech.com*
6. Low-Power Electronics Design, edited by C. Piguet, CRC Press, 2005.
7. International Technology Roadmap for Semiconductors, http://public.itrs.net
8. K.W. Lee, SoC R&D Trend for Future Digital Life, *IEEE Asia-Pacific Conf. on Advanced System Integrated Circuits*, pp. 10–13, 2004.
9. M. Ruggiero, F. Angiolini, F. Poletti, D. Bertozzi, L. Benini, R. Zafalon, Scalability Analysis of Evolving SoC Interconnect Protocols, *Int. Symposium on System-on-Chip*, 2004.
10. W. Wolf, *The Future of Multiprocessor Systems-on-Chip, Design Automation Conference*, pp. 681–685, 2004.
11. S3C2410, System LSI, *www.samsung.com*
12. ARM9E Family, *www.arm.com/products/CPUs/families/ARM9EFamily.html*
13. Intel PXA27x Processor Family, *www.intel.com/design/pca/prodbref/253820.html*
14. OMAP5910, *http://focus.ti.com/docs/prod/folders/print/omap5910.html*
15. Nomadik, *www.st.com/stonline/prodpres/dedicate/proc/proc.htm*
16. i.MX21 Applications Processor, *www.freescale.com*
17. V.J. Falgiano, W.R. Newberry, Entry level MCM/MCP trade-offs and considerations, *Electronic Components and Technology Conf.*, pp. 796–804, 1994.
18. M. Yokotsuka, Memory Motivates Cell-Phone Growth, *Wireless Systems Design Magazine*, April 2004.
19. Key Trends in the Cellular Phone Market for Flash Technology, *SEMICO Research Corporation, analyst R. Wawrzyniak, consultabile su www.intel.com*
20. D.K. Schulz, Memory Issues in SoC, *Int. Seminar on Application-Specific Multi-Processor Soc MPSoC'04, consultabile su http://tima.imag.fr/MPSOC/2004/slides/dks.pdf*
21. L.R. Zheng, M. Shen, H. Tenhunen, System-on-Chip or System-on-Package: Can we Make an Accurate Decision on System Implementation in an Early Design Phase? *Symposium on Mixed-Signal Design*, pp. 1–4, 2003.
22. J. Rasmusson, F. Dahlgren, H. Gustafsson, T. Nilsson, Multimedia in Mobile Phones—The Ongoing Revolution, *Ericsson Review* no. 02, 2004.

23. J. Jang, S. Lim, M. Oh, Technology Development and Production of Flat Panel Displays in Korea, *Proceedings of the IEEE*, Vol.90, no. 4, pp. 501–513, 2002.
24. T. Ojanperä, R. Prasad, Wideband CDMA for Third Generation Mobile Communications, *Artech House Publishers*, 1998.
25. L. Benini, G. De Micheli, Networks on Chips: a New SoC Paradigm, IEEE Computer, Vol. 35, no. 1, pp. 70–78, 2002.
26. J. Ayala, M. L. Vallejo, D. Bertozzi, L. Benini, State-of-the-art SoC Communication Architectures, *in Embedded Systems Handbook*, CRC Press, 2004.

Chapter 2
Nonvolatile Memories:
NOR vs. NAND Architectures

L. Crippa, R. Micheloni, I. Motta and M. Sangalli

2.1 Introduction

Flash memories are nonvolatile memories, i.e., they are able to retain information even if the power supply is switched off. These memories are characterized by the fact that the erase operation (the writing of logic "1") has to be performed at the same time on a group of cells called a sector or block; on the other hand, the program operation (the writing of logic "0") is a selective operation during which a single cell is programmed. The fact that the erase can be executed only on an entire sector allows one to design the matrix in a compact shape and therefore in a very competitive size, from an economic point of view. Depending on how the cells are organized in the matrix, it is possible to distinguish between NAND Flash memories and NOR Flash memories. The main electric characteristics are reported below.

For the Table 2.1 the following definitions hold:

- Dword: 32 bits;
- Output parallelism: the number of bits that the memory is able to transfer to the output at the same time;
- Data read/programmed in parallel: the number of addressable bits at the same time during read/program operation;
- Read access time: time needed to execute a read operation, excluding the time to transfer the read data to output.

L. Crippa
Qimonda Design Center, Vimercate, Italy

R. Micheloni
Qimonda Design Center, Vimercate, Italy

I. Motta
Numonyx, Agrate Brianza, Italy

M. Sangalli
Qimonda Design Center, Vimercate, Italy

R. Micheloni et al. (eds.), *Memories in Wireless Systems*,
© Springer-Verlag Berlin Heidelberg 2008

Table 2.1 Comparison between NOR and NAND Flash memories

	NOR	NAND
Memory size	$<= 512$ Mbit	1–8 Gbit
Sector size	$\sim$1 Mbit	$\sim$1 Mbit
Output parallelism	Byte/Word/Dword	Byte/Word
Read parallelism	8–16 Word	2 Kbyte
Write parallelism	8–16 Word	2 Kbyte
Read access time	<80 ns	20 μs
Program Time	9 μs/Word	400 μs/page
Erase time	1 s/sector	1 ms/sector

The aim of this chapter is to explain the reasons why the performances in read program and erase are so different owing to the connection used to create the memory matrix.

2.2 The Read Operation in Flash Memories

One of the most important parameters for any kind of memory, not only for Flash memories, is the access time to the data stored in it.

The access time is the temporal interval that passes through any address commutation to the moment in which the addressed data are available to the output. This time is commonly defined as "asynchronous access time." The term "asynchronous" is used to distinguish this kind of read operation from another one, defined as "synchronous" and characterized by the fact that the read data should be synchronized to the output with an external clock.

Nowadays, the most commonly used architectures for Flash memories are called NOR and NAND. The common element of both architectures is the nonvolatile (single) cell. The program operation acts on the threshold voltage of the Flash cell, modulating its value and the current/voltage characteristic as a consequence.

The read of a nonvolatile memory cell is done applying convenient voltages to its terminals and measuring the current that flows into the cell. NOR and NAND memories measure this current in different ways. In the following, the measuring method will be discussed separately to better highlight the fundamental differences.

2.2.1 NOR Architecture

In NOR Flash memories, the read of a matrix cell is done in a differential way, i.e., making a comparison between the current of the read cell and the current of a reference cell which is physically identical to the matrix cell and biased with the same voltages V_{GS} and V_{DS}.

Fig. 2.1 Current/voltage characteristics as a function of the threshold voltage of a Flash cell

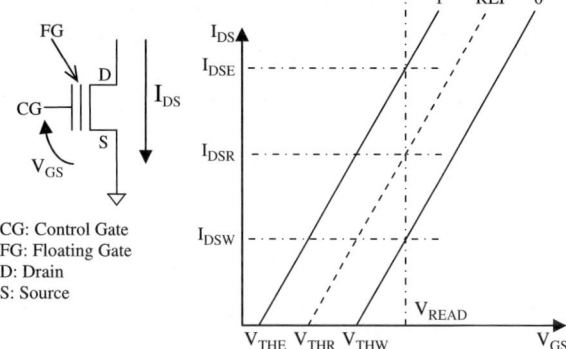

CG: Control Gate
FG: Floating Gate
D: Drain
S: Source

In the case of memories storing only one bit per cell, the electrical characteristics of the I_{DS}-V_{GS} of the written cell (logic "0") and of the erased cell (logic "1") are separated as sketched in Fig. 2.1; this is due to the fact that the two cells have different threshold voltages V_{THW} and V_{THE}.

Since the read voltage V_{READ} applied to the control gate is the same, the "written" cell (logic "0") sinks a current I_{DSW} lower than the current sunk by the "erased" cell (logic "1") I_{DSE}. To distinguish correctly the two characteristics it is necessary to act on the threshold voltage of the reference cell so that its characteristic is placed between the erased and the written cell characteristics. In this way, if the same V_{GS} is applied to all cells (V_{READ}, in Fig. 2.1), the cell will be recognized as erased if its current is higher with respect to the current of the reference cell I_{DSR}; vice versa, it will be seen as written. The current of the cells is converted into a voltage by means of a current to voltage converter (I/V), a circuit able to supply to the output a voltage whose value depends on the value of the current at its input. The voltages are then compared through a voltage comparator which gives at its output the cell status: the logic levels "0" or "1" (Fig. 2.2).

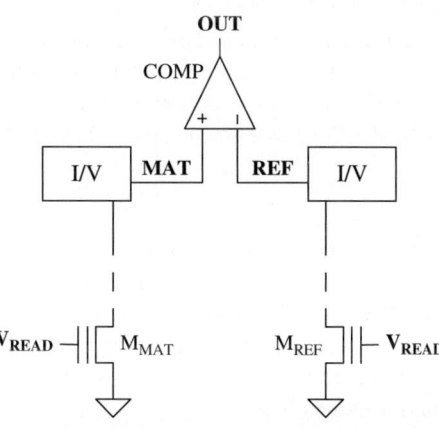

Fig. 2.2 Block scheme of a circuit used to compare two currents

It is important to avoid the drain of the cells from reaching too high a voltage during the read operation because it may cause a spurious program operation, thus modifying the threshold voltage of the cell. For this reason the drain has to be biased with a voltage lower or equal to 1 V. Therefore, a circuit able to fix the drain voltage at 1 V is needed before the I/V converter.

In most cases the circuitry necessary to execute a read operation (commonly known as *sense amplifier*) is composed of the following fundamental blocks:

– I/V converter;
– drain voltage limiter (this voltage is usually about 1 V);
– output comparator.

One of the simplest circuit implementations of a sense amplifier is sketched in Fig. 2.3: on the left hand side there is the matrix cell (M_{MAT}), while on the other side there is the reference cell (M_{REF}). The same voltage (V_{READ}) is applied to the gates of both cells, while on their drains the voltage limiter ($V_{DS}REG$) is placed in order to limit the drain voltage. The voltage limiter is designed using a NMOS transistor and biasing its gate with a fixed voltage. The I/V converter is realized using a resistive load, and its output nodes (MAT and REF) are compared by the voltage comparator COMP.

Over the years, new and more complex circuits have been realized to improve the behaviour of the sense amplifier and to reduce the access time.

For example, the drain regulator is designed using closed loop structures, while the I/V converter is realized using current mirrors or active loads [1].

It is possible to have a sensible improvement using the so-called equalization technique: before the comparison takes place, the nodes MAT and REF are

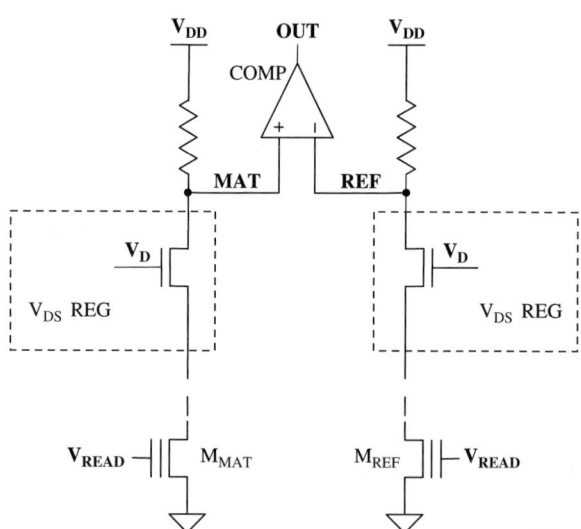

Fig. 2.3 Basic scheme of a sense amplifier

"equalized" (i.e., they are forced) to the same value. In this way the node MAT is always near to the final voltage value.

A further equalization technique can be used to increase the read speed. This kind of equalization, that seems simple from a circuital point of view, consists in using a number of reference cells equal to the number of cells which are read at the same time [2]. To gain a real benefit from this equalization method, it is very important for the reference cell to have the same load condition as the matrix cell. For this reason, the capacitive load of the matrix cell is recreated on the drain of the reference cell; a real bitline is usually used to have a complete matching.

Using the above described architecture, it is possible to read in a dynamic differential way, obtaining a correct separation in voltage of nodes MAT and REF, as soon as the equalization phase is finished: being equal to the sensibility of comparator COMP, a faster read is performed than with the classic approach, where it is necessary to wait until the nodes MAT and REF are stable.

In Fig. 2.4 the evolution of nodes MAT and REF is shown in the case of a simple differential read operation and in the case of a dynamic differential read operation. Typically, during the read operation, the time dedicated to the comparison of the currents is about 10–20 ns.

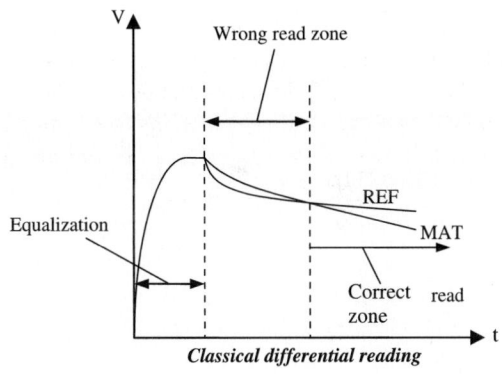

Classical differential reading

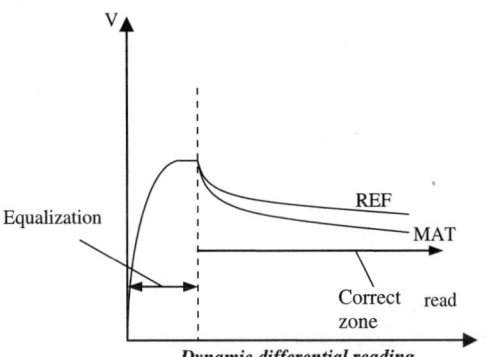

Fig. 2.4 MAT and REF node transients in classical and dynamical differential reading

Dynamic differential reading

2.2.2 NAND Architecture

In the NAND Flash architecture, the cells are connected in series, in groups of 16 or 32. Two selection transistors are placed at the edges of the stack, to ensure the connections to ground (through M_{SSL}) and to the bitline (through M_{DSL}).

This basic structure is shown in Fig. 2.5. When a cell is read, its gate is set to 0 V, while the other gates of the stack are biased with a high voltage (typically 4–5 V), so that they work as pass-transistor, regardless of their threshold voltage.

An erased NAND Flash cell has a negative threshold voltage; on the contrary, a programmed cell has a positive threshold voltage but, in any case, less than 4 V. In practice, driving the selected gate with 0 V, the series of all the cells will sink current if the addressed cell is erased, otherwise no current is sunk if the cell is programmed.

Figure 2.6 shows the threshold voltage distributions V_{TH} for the erased and programmed memory cells; note that, for gate voltages above the right margin of the programmed distribution (V_{THSD}), the cells always sink current whatever their threshold voltage is. This feature is used particularly when the cell should operate as a pass-transistor.

Unlike NOR Flash memory, the current to be sensed in these serial structures is very low. This value of current is typically 200–300 nA (tens of μA in NOR architecture). It is unfeasible to detect such current with a differential structure as in the previous section.

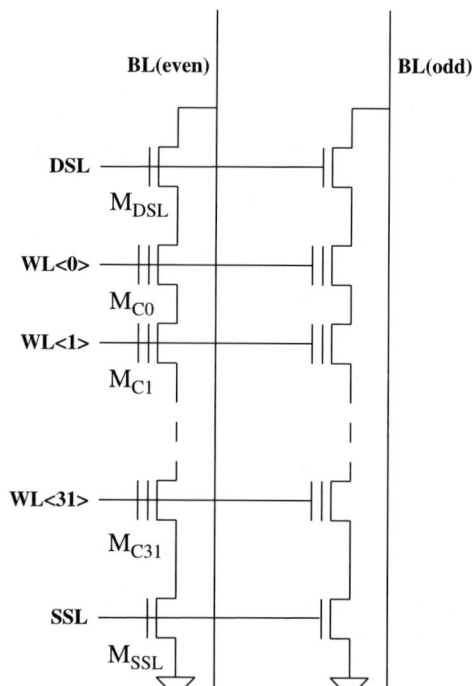

Fig. 2.5 Matrix structure in NAND architecture

Fig. 2.6 Threshold voltage distributions for erased and programmed cells

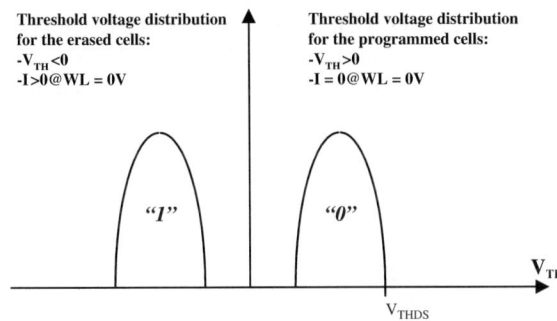

The reading method in NAND memories is the charge integration, which uses the parasitic capacity of the bitline. This capacitance is precharged to a fixed value (typically 1.2 V): if the cell is erased, it sinks current and discharges the bitline; otherwise, if it is programmed, it does not sink current and the bitline keeps its initial value. There are many circuits to detect the charge status of the bitline parasitic capacitance: they can be summarized with the structure shown in Fig. 2.7a. The bitline parasitic capacitance is indicated with C_{BL}; the electric characteristic of the NAND string is summarized with a current generator (I_{CELL}).

During the bitline precharge, the gate terminal of M_P is kept at GND (0 V), while the gate of M_N is at a fixed value V_1 (for example, 2 V). At the end of the charge transient, the voltage V_{BL} on the bitline is:

$$V_{BL} = V_1 - V_{THN} \tag{2.1}$$

where V_{THN} represents the threshold voltage of the n-channel transistor M_N.

The bitline precharge phase usually lasts 2–6 µs, in order to reduce the current consumption peak from the power supply VDD. The voltage V_{OUT} is initially precharged at VDD. After the precharge phase, M_N and M_P are turned off, leaving OUT and BL nodes floating (high-Z status), i.e., charged to their precharge value.

After the precharge phase, the "evaluation phase" begins, where the cell current is checked. If the cell does not sink current, the bitline capacitance keeps its precharged value; if the cell sinks current, the bitline begins to discharge. At the end of T_{VAL}

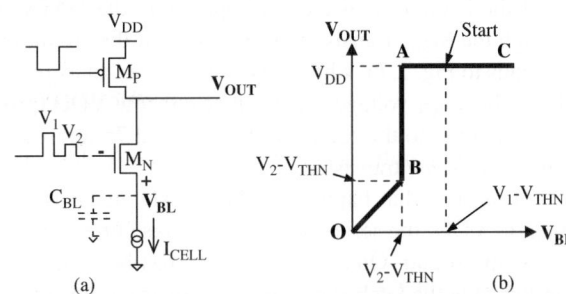

Fig. 2.7 (a) Structure to detect the bitline discharge and (b) V_{OUT} characteristic as a function of V_{BL}

at the thermal equilibrium in the silicon lattice. The greater part of the generation of hot electrons takes place in the region where the electric field is more intensive, i.e., the depletion region near the drain.

In this condition some electrons acquire enough energy (greater than 3.1eV) to overcome the potential barrier at the channel-oxide interface.

Applying a voltage to the control gate, a transversal electric field is then created in order to support the injection of electrons from the channel to the oxide and their gathering into the floating gate.

In any case, the injection of electrons comes naturally to an end, because the increasing of the negative charge in the floating gate causes a continuous decrease in the gate potential, and therefore the electrons are less and less attracted.

The program operation through hot electrons is a fast operation, but the current necessary for the program to take place is very high. Obviously, the greater the number of cells to be programmed at the same time, the greater is the current consumption. For example, to program 64 cells, it is necessary to produce on chip 3.2 mA, supposing that the current consumption of a single cell is about 50 μA. Therefore the programming of a huge number of cells using this kind of mechanism becomes an expensive operation, especially concerning the area required to design the peripheral circuits needed to create such high currents.

The voltages applied to the cell during the program operation are:

- 4.5 V on drain;
- 9 V on gate;
- 0 V on body and source.

Actually, the voltages applied to the cell terminals depend on the technological node. The cell should have a good speed during the program operation and a good reliability as regards parasitic effects such as the drain turn on, the snap back, the soft programming during the read operation, and the soft erasing during the erase operation. Therefore, in the choice of voltages to be applied, it is necessary to consider aspects as the channel length, the program efficiency of the cell, the drain current, and the process variation. Moreover it is very important to apply the voltages with high precision; time conditions being equal, a variation in the value of the voltages applied during the program operation may cause a variation in the drain current and therefore a variation in the cell's capacity to accumulate the electrons.

Last but not least, it is very important to follow a precise sequence to bias the cell terminals; once the voltage has been applied to the gate of the cell, the drain will be biased. This choice depends on the cell matrix structure; particularly sensible cells with 4.5 V on drain and 0 V on gate might present snap back effects or undesired program/erase effects.

A distribution of threshold voltage is obtained as result of a modify operation (program or erase operation) executed on a number of cells. This is due to different factors such as process variation, dissymmetrical geometry, power supply variation, and source and drain modulation. Despite all these effects, it is important that the distributions have a precise and well controlled width so that they can be correctly allocated in the working window of the memory cells. This requirement has been highlighted with the introduction of multilevel memories, i.e., memories in which there is more than one bit per cell. As the number of bits to be stored in the memory cell increases, the number of distributions to be allocated in the working window increases as well. In fact, due to technological and reliability constraints, usually the designer cannot enlarge the working window. For example, if there are three bits to store, than eight distributions must be allocated.

The width of the distributions may be controlled by choosing an appropriate program algorithm. For example a *program and verify* approach can be chosen, where the program pulse is followed by a verify operation. The verify operation consists in controlling the cells under program if they have reached the target threshold voltage. In the NOR memories, the threshold voltage of the memory cell is compared to the threshold voltage of a reference cell, as in the read operation. But there are two factors that make the verify operation different from the read operation: the reference cell used, and the time necessary to execute the comparison. Usually all the cells are overprogrammed with respect to the read reference voltage, so that there is a margin during the read operation. Obviously it is not possible to take great margins, especially when many distributions must be allocated in the same working window; therefore, in order to guarantee a greater precision to the verify operation, the timing is relaxed compared to the one used during the read operation. The cells that result as programmed after the verify operations (i.e., that reach the desired threshold voltage) are left in that position, while another program pulse is applied to the other cells. Either when all the cells are programmed, or when all the attempts to program the cells have been made, the algorithm finishes. While in the first case the operation finishes with success, in the other case the "fail" information is communicated to the user.

If a cell should not be programmed, it is necessary to avoid the applying of high voltages on gate and drain at the same time. Due to the memory architecture, there will be some cells with a high voltage on the gate but 0 V on drain, and some other cells with a high voltage on drain but 0 V on gate. These cells may suffer from different disturbances. In Fig. 2.9, a little matrix is sketched where the potentials for the cell under program and for the cells sharing the same bitline and the same wordline are stressed. The cells sharing the same bitline suffer the so called "drain disturb"—owing to the potential applied on the drain, the already programmed cells (therefore with a negative charge collected in the floating gate) may show a loss of charge. On the contrary, the erased cells sharing the same wordline may suffer a program operation by Fowler–Nordheim tunnelling. Therefore, they can show, at the end of the program operation, a greater threshold voltage. For the programmed cells sharing the wordline with the cell under program, an injection of electrons from the floating gate to the control gate may take place, and these cells will show, at the end, a lower threshold voltage.

Fig. 2.9 A NOR memory architecture showing the biasing of wordlines and bitlines during the program operation; the cell under program is the highlighted one

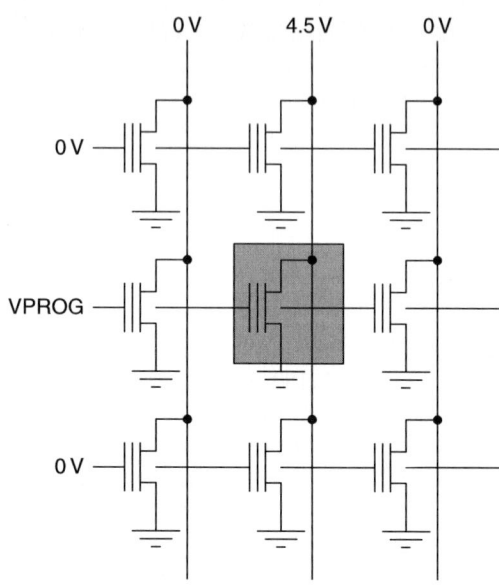

2.3.2 Program in NAND Memories

As already mentioned above, the program operation in NAND memories takes place thanks to a different physical principle: it is exploiting the quantum effects of tunnelling of electrons in the presence of a high electric field. In particular, the operation depends on the polarity of the electric field: if it is directed from substrate to gate, than a program of the cell is obtained; an erase operation occurs if the polarity of the electric field is the opposite one.

In reality the tunnelling effects may be two:

– the *channel tunnelling*, where the electric field is applied between gate and substrate, while the drain and source terminals of the cell are floating;
– the *junction tunnelling*, where the electric field is applied between the gate and one of the two terminals (drain or source).

The effect used in NAND memories is the channel tunnelling. In Fig. 2.10 it is possible to see the modifications of the potential barrier: the potential barrier of the oxide insulating the floating gate from the substrate during the read operation, and how it is modified during the program operation due to the intensive electric field applied.

During programming the number of electrons that pass through the tunnel oxide depends on the electric field: the greater the electric field, the greater is the probability of the injection of electrons. In order to improve the program performance it is necessary to have high electric fields and high voltages. This requirement is one

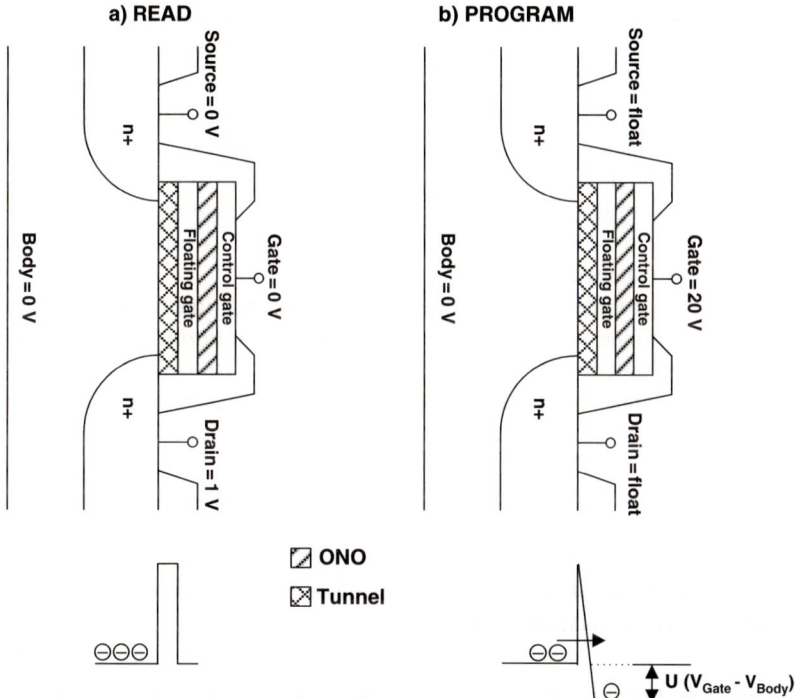

Fig. 2.10 Biasing voltages for a NAND cell memory in case of (**a**) a reading operation, and (**b**) a programming operation and relative band diagram

of the main disadvantages of this method of programming, because damages of the tunnel oxide are due to these high voltages.

A reduction of the thickness of the dielectric seems to be the best solution to this problem. In fact, in this way, the injection efficiency improves, while the voltages necessary to obtain the program and the erase of the cells (and therefore the total energy of electrons crossing the oxide) decrease. Unfortunately, if the dielectric is too thin other negative effects may take place, such as the *stress induced leakage current* (SILC).

Another disadvantage of the tunnelling method for programming is the time required, which is typically longer than in the channel hot electrons case.

On the other hand, the main advantage is the current required for this operation, which is rather contained (on the order of a nanoAmpere per cell). This characteristic makes the Fowler-Nordheim method the right one to program in parallel a great number of cells.

As with the cells of a NOR memory, the algorithm used to program the cells in a NAND memory is a *program and verify* algorithm (after a program pulse, a verification on the threshold voltage of the cell is done). In the NAND memories, as in NOR ones, the verify threshold voltage used is higher than the one used during

the voltages and their timings must be chosen with care, because no erase failure is admitted either at the beginning or during the device lifetime (100 k program/erase cycles allowed, 10 years lifetime).

2.4.2 Erase in NAND Architecture

In NAND architecture Flash memory the erase is performed by biasing the IP-well with a high voltage and keeping to GND the wordlines of the sector to be erased. Why? To generate negative voltages, negative boosters and high-voltage triple-well transistors are needed [10]; these transistors would be needed also to bias the row decoder with a negative voltage, and the row decoder itself should have at least one of these transistors to transfer the negative voltage to the wordlines. On the other hand, to avoid the use of negative voltages means saving from the point of view of the masks and lithographic complexity of technology. Also, in NOR architecture it is not trivial to place a row decoder able to drive negative voltages in the wordline Y-pitch, because this circuitry is composed at least of three transistors [11]; since in NAND architecture the Y-pitch is further reduced, it is more convenient to design a single N-channel transistor row decoding to pass positive (or zero) voltage values. Furthermore, as we will explain later, it is also possible to get the information we need on the position of the erased distribution with sufficient precision using positive voltages only: even from this point of view, it is not necessary to push towards the use of negative voltages.

In NAND architecture, the erase pulse is applied to the bulk terminal, which must be brought to a voltage higher than in the NOR case (the common source node is left floating). This terminal is shared between all the blocks to get a more compact matrix and to reduce the number of the structures to bias the iP-well (in NOR architecture there is one of these structures for each sector or group of sectors). By contrast, the parasitic capacitance to load is much higher, and the sectors not to be erased (i.e., all except one) should be managed properly to avoid spurious erase.

The advantage of NAND architecture from the erase point of view is that it is possible to locate the erased distribution into the negative threshold half-plane, i.e., the erased distribution may be depleted. Unlike NOR architecture, the cell threshold may be brought to the negative because the cells must act as pass-transistors, i.e., biased in conduction, for the reading mechanism to work. The subthreshold current does not represent a leakage contribution, because the selection transistors of the unselected bitlines prevent them from injecting any spurious current; for this reason, it is not necessary to bias the unselected wordlines with negative voltages to switch them off. The erased distribution width is huge, but without great effect on the series resistance of the stack when it has to be read, because during read algorithm all the cells of the stack except the addressed one are biased with a sufficiently high gate voltage (V_{PASS}, about 4.5 V).

In NAND architecture a great precision in placing the erased distribution is not necessary; all that is needed is an adequate margin with respect to the read condition.

For this reason, the erase is performed with a single impulse, calibrated to bring all the erased distribution to the negative, followed by a verification phase. The NAND specification helps in this case, allowing the management of bad blocks; if the block after an erase pulse does not meet with the erase verify, the user must consider it failed and store this information so as to prevent the use of this block. (If the malfunction occurs during the factory test, the sector can be directly marked as bad. The user is required to check all the blocks to avoid using the bad ones.)

The way the erase verify is performed in NAND architecture is substantially different from that used in NOR architecture. It is not possible to know exactly where the edges of the erased distribution are, since negative voltages are not provided on-chip. On the other hand, what is important is that all the cells of the string after erase are read as erased, i.e., they must be read as "1" when their gate is biased to GND and the other cells of the string are biased as pass-transistors. This condition will apply to all the cells of the string. The erase verify in NAND architecture is therefore a "stack operation," because it requires that the whole stack is read as "1" when it is biased with GND (i.e., when biasing all the wordlines of the stack with such voltage). As already said, the exact placement of the erased distribution is unknown, but much time is saved for the erase verify, since it is possible to have the required information with only one read operation, instead of as many reads as the cells of the stack.

A further requirement is that the erased distribution have enough margin to contain the degradation due to subsequent erase and program cycles. Figure 2.17 is a qualitative representation of the "cycling window" of a NAND cell as a function of the number of program/erase cycles (program and erase are executed in a "blind" way, i.e., applying the program and erase pulses without verification). The thresholds of both the erased and the programmed cells increase with the number of cycles. This phenomenon is due to gain degradation and charge trapped into the oxide. The macroscopic effect on the erase operation is such that the erase pulse that allows having pass result at erase verify at the beginning of the operating life could no longer be enough as the cycle number increases.

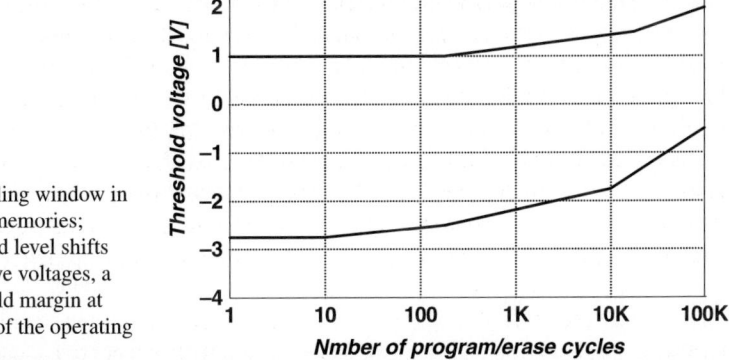

Fig. 2.17 Cycling window in NAND Flash memories; since the erased level shifts towards positive voltages, a proper threshold margin at the beginning of the operating life is required

The cycling degradation also affects the NOR architecture, but in this case it is possible to apply further erase pulses if they are set at the beginning of operating life. In other words, in this case the cycling degradation results in an erase time increase (both for the erase time itself and for the soft-programming phase), but the NOR specifications allow this. In NAND architecture instead, the specifics leave no room for a further erase pulse if the first is not effective; for this reason, the erase pulse level and width must be carefully calibrated after accurate analyses at the process level.

Also in NAND architecture, a preconditioning before the erase pulse may be useful to get more uniform threshold voltages so as to reduce the distribution spread. Thanks to the program mechanism, which uses the tunneling phenomenon, a lot of time is saved if the preconditioning is executed at a time with a unique pulse on the wordlines of the block to be erased.

The erase sequence in NAND architecture is shown in Fig. 2.18, together with the effect that the various phases have on the distribution. The preconditioning is not displayed because it is often not used (in any case, its effect is similar to that in NOR architecture).

The nodes involved in the erase operation must be discharged carefully in NAND architecture also. Since the bitlines are floating, they are charged to a voltage potential, depending on the bulk voltage and the capacitive loads. The bulk discharge must be well controlled, as in NOR architecture. The common source node may be initially left floating (it starts discharging anyway, due to its capacitive coupling to the bulk), and then connected to ground when the bulk discharge is over. The same concept applies also to the bitlines: after the initial discharge due to their coupling with the bulk, they are discharged to ground, thanks to proper transistors.

The selected wordlines are already kept to ground, but the unselected are floating, so they are discharged to ground by activating all the row decoders. Great attention must be paid at the technological and manufacturing levels to the electrical characteristics of the row decoder. The leakage should be as low as possible. In fact, during the erase pulse the wordlines of the unselected blocks are left floating, so they are free to load to a certain voltage by their capacitive coupling with the bulk, and tunneling is not allowed. However, if there is a leakage at the row decoder level, the unselected wordlines may discharge; if this happens, the voltage drop across the tunnel oxide may be enough to trigger a spurious erase of the unselected blocks.

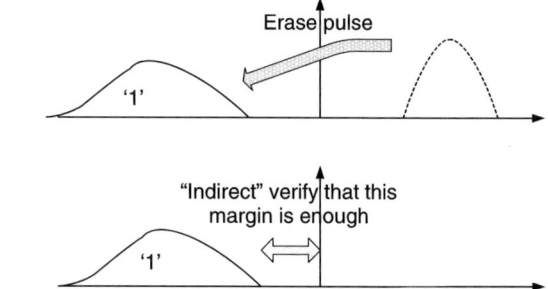

Fig. 2.18 Erase algorithm phases for NAND architecture Flash memory and their effect on the distribution

In summary, in a NAND Flash the typical block erase lasts approximately 1 ms; 800 µs for the erase pulse and about 100 µs for the erase verify. Usually, the preconditioning is not carried out, but if it is carried out it lasts no more than 100 µs. The reader should note that in the NAND architecture the erase time of a block is really independent of the size of the sector, because none of the operations that compose the algorithm is made at the page level, but everything is done at the block level. Furthermore, the erase time is "rigid," as all the phases last exactly the same time for all the blocks, since "repetitions" are not possible to recover the failed blocks.

References

1. Campardo G, Micheloni R, Novosel D (2005) VLSI-design of nonvolatile memories. Springer series in advanced microelectronics
2. Elmhurst D et al. (2003) A 1.8V 128Mb 125MHz multi-level cell flash memory with flexible read while write. ISSCC Dig Tech Pap 286–287
3. Jung TS (1996) A 3,3-V 128-Mb multilevel NAND flash memory for mass storage applications. ISSCC Dig Tech Pap 32–33
4. Lenzlinger M, Show EH (1969) Fowler-Nordheim tunnelling into thermally grown SiO_2. IEDM Tech Dig 40:273–283
5. Hu C (1993) Future CMOS scaling and reliability. P IEEE 81:682–689
6. Kenney S et al. (1992) Complete transient simulation of flash EEPROM devices. IEEE T Electron Dev 39:2750–2757
7. Bez R et al. (2003) Introduction to flash memory. P IEEE 91:554–568
8. Cappelletti P et al. (eds) (1999) Flash memories. Kluwer, Norwell, MA
9. Pavan P, Bez R, Olivo P, Zanoni E (1997) Flash memory cells—an overview. P IEEE 85: 1248–1271
10. Umezawa A et al. (1992) A 5 V-only operation 0.6-µm flash EEPROM with row decoder scheme in triple-well structure. IEEE J Solid-St Circ 27:1540–1546
11. Motta I, Ragone G, Khouri O, Torelli G, Micheloni R (2003) High-voltage management in single-supply CHE NOR-type flash memories. P IEEE 91:554–568

Chapter 3
Nonvolatile Memories: Novel Concepts and Emerging Technologies

R. Bez, P. Cappelletti, G. Casagrande and A. Pirovano

3.1 Introduction

In the last decade the portable systems market (palmtop, mobile PC, mp3 audio player, digital camera, and so on) has been characterized by an impressive growth in term of sold units and revenues, attracting more and more interest of the semiconductor industries on nonvolatile memory (NVM) technologies. In fact, these technologies have demonstrated the ability to accomplish the requirements of some portable applications to store and execute embedded code, as well as to allow for mass storage applications with good performance and very low costs. The NVM market has thus seen a fast growth, rising from $2B of revenues in 1998 up to $14B in 2004. A large part of this increase has been related to the huge demand for memory devices with larger storage capability, better performance, reduced power consumption, smaller form factors, and lower weights and costs.

For more than fifteen years Flash memory floating-gate (FG) technology has been able to follow the evolution of the semiconductor roadmap [1, 2]. Recent technological developments have definitely confirmed the division of this market into two mainstreams based on different variations of the Flash technology concept. Because of their multilevel capabilities (MLC) and their fast random access for execution in place (XiP) applications, NOR memories are a cost effective solution to the challenge of stacking code and large amounts of data into a single chip. With a cell size close to the minimum lithographically defined feature of $4F^2$ and MLC capabilities, NAND technology is expected to be the lowest cost NVM and the better solution

R. Bez
Numonyx, Agrate Brianza, Italy

P. Cappelletti
Numonyx, Agrate Brianza, Italy

G. Casagrande
Numonyx, Agrate Brianza, Italy

A. Pirovano
Numonyx, Agrate Brianza, Italy

R. Micheloni et al. (eds.), *Memories in Wireless Systems*,
© Springer-Verlag Berlin Heidelberg 2008

for high-density data storage for the rest of this decade. Although the current visibility allows us to predict a scenario where the floating-gate concept is a valuable solution till the arrival of the 45 nm technology node, there are physical limitations to be addressed and the striking trade-off between oxide thickness and data retention should be addressed with alternative solutions. In this scenario, emerging NVM concepts can try to exploit their performance advantages and better scaling capabilities in order to enter the actual Flash-centric memory market and eventually become the leading NVM technology. Boosted by this perspective and by the fast growth of the NVM market, in recent years more than 30 alternative NVM concepts have been proposed to replace Flash memory and some of them have also been claimed to be capable of replacing DRAM, becoming potentially a unified memory concept.

Despite the fact that the proposed emerging memory concepts are characterized by better performance compared to Flash technology, up to now none of these alternative solutions has been widely exploited for commercial products. In fact, there are still several issues, mainly from the technological point of view, that must be addressed to make these concepts really competitive with the existing NVM mainstream. The aim of this chapter is to present the current challenges and limitations of Flash technology in the oncoming years, discussing the potentialities of the emerging NVM concepts and providing a complete overview of the NVM market in the next decade. Although applications that effectively exploit new NVM concepts are still lacking, a deep comprehension of the development of the NVM market and how emerging concepts and novel memory technologies could offset the current market is a strategic key point to foresee novel opportunities and to anticipate the development of new applications.

3.2 Flash Technologies: Challenges and Perspectives

Looking back over the evolution of the silicon era, the demand for memory devices capable of supporting CMOS logic has been one of the most striking issues for the semiconductor industries. However, the ideal storage system has never been discovered. Several concepts and technologies have been exploited to patch this shortcoming, none of them able to satisfy all the requirements at the same time. The development of memory technologies has thus been driven by the growth of applications demanding a specific memory concept that could better satisfy their requirements. As a proof of this, the most relevant phenomenon of this past decade in the field of semiconductor memories has been the explosive growth of the NVM market, driven by cellular phones and other types of electronic portable equipment and, in recent years, the equally explosive growth of the Flash NAND market due to the increasing demand for data storage capabilities. The development of portable applications has thus promoted the leading role of Flash technology in the NVM market, actually divided into two mainstream technological solutions, namely NOR and NAND Flash technology.

The first developments in NOR Flash technology date back to the mid 1980s [3, 4] and the success in the NVM market as EPROM replacement happened some years after the introduction of the first Flash product in 1988. The increasing demand for different applications and the fast technological development soon allowed NOR technology to promptly respond to the various needs of the market, becoming the leading NVM technology for all the 1990s. The ability to store several levels in each cell (corresponding to more than one bit of information for each cell) and the very fast random reading access suitable for code execution (XiP—Execution in Place) made the NOR technology a cost-competitive solution to store both data and code on a single chip. NOR Flash units are today available in several sizes (from 1 Mbit up to 512 Mbit), power supply voltages (from 1.8V to 5V), reading parallelism (serial or random access x8, x16, x32, burst, or pages), and memory partitioning (bulk or sector erasing, equal sectors or boot block sectors schemes). This wide range of products has been developed to satisfy the requirements of specific applications and demonstrates the extreme versatility of NOR Flash technology. A wide application spectrum, excellent performance/price ratio, and outstanding reliability are the main factors at the base of the impressive success obtained by this technology in the last years.

NAND Flash technology was initially developed in the mid 1980s too [5], but the time required to get the technological maturity, and most of all the development of a suitable market that could exploit the potentialities of this technology required a longer time with respect to NOR Flash. Because of the extremely small cell size, very close to the minimum theoretical size of $4F^2$, where F is the lithographic node dimension, and because of its ability to store more than one bit per cell (multilevel capability), NAND technology is today the cheapest solution for storing large amounts of data with very high densities and competitive costs. Although the quite long random access times, the lower reliability requirements, and the corresponding need to adopt error correction codes make this technology less versatile than NOR Flash, the very low cost and the fast writing throughput are at the base of the success of NAND Flash for all mass storage applications.

The demand for an NVM product is thus strictly related to the appearance of a driving application that can exploit its potentialities. However, the acquisition of a leading competitive position based on performance advantages cannot be sustained without the ability of memory devices to scale down and reduce their costs. From this perspective, although in the next years it is expected that portable equipment will require more and more NVM devices, the capability to catch these opportunities is strictly related to the scalability of Flash technology in order to provide higher storage capabilities at lower costs.

Throughout the history of the development of Flash, the cell size has been reduced through a combination of lithography reduction and self-alignment techniques [6]. This trend will continue in next-generation devices [7], with the increasing importance of self-alignment techniques to compensate for the difficulties of scaling down the lithographic registration dimensions. But, as Flash scales down into the sub-45 nm regime, challenges arise from physical limitations related to the high voltages needed by the programming and erase mechanisms, the stringent charge

The storage media can be a complete system that is a well defined product usable as a storage media device, or a subsystem of a more complex system. The second mentioned typology of storage media won't be treated in this chapter; in the second case the storage media will be present in any more complex system as a memorization subsystem. The main parts of a storage media device are the memory, the memory controller, and the firmware that manages the communication between the different parts and the external world.

4.1.2 The Flash Memory Card As a "Removable Storage Medium"

The Flash memory cards are to all intents and purposes, storage media systems, specifically, systems that are removable. At the present time, such products are coming to be embedded in different devices, and hence are known as embedded storage media devices. Because they are designed to store data, they are based on a nonvolatile memory. It is important to underline that it is not trivial to "transform" a Flash memory into a Flash card. First of all, a Flash card embeds at least two integrated circuits, the Flash memory and the controller; the controller runs the firmware that manages communications between the Flash memory and the external world. Recognizing a Flash card as a storage media device, we will now describe its main differences and added values compared to a simple memory device:

- the external protocol is independent from the memory communication protocol;
- the external world can communicate with the memory communication protocol by means of a higher level language;
- the embedded controller makes the Flash card, in theory, independent from both the memory features and the memory technology;
- the Flash card implements by itself the memory access procedures to improve the device's reliability.

4.1.3 A Flash Memory Card

Now we can go into more detail about the Flash card, the world with which it interacts, and its composition. Figure 4.1 shows the form factors of different Flash cards currently on the market.

At first glance the Flash card appears to be the heir to the floppy disk, and this is quite true for the user, even if the dimensions and the storage capacities are quite increased. The main and substantial difference between a floppy disk and a Flash card consists in the design and manufacture of each. The world that interacts with the Flash card expects that the device works like a floppy disk—that is, the *host* doesn't want to deal with internal device implementation details, it needs only to know that it can use the device to write some information, that the information doesn't get lost, and that if needed it can be retrieved. Typically the world that interacts with the Flash memory card is known as the *host*—that is, any object capable of communicating with the card by means of a communication protocol. However, to explain the Flash

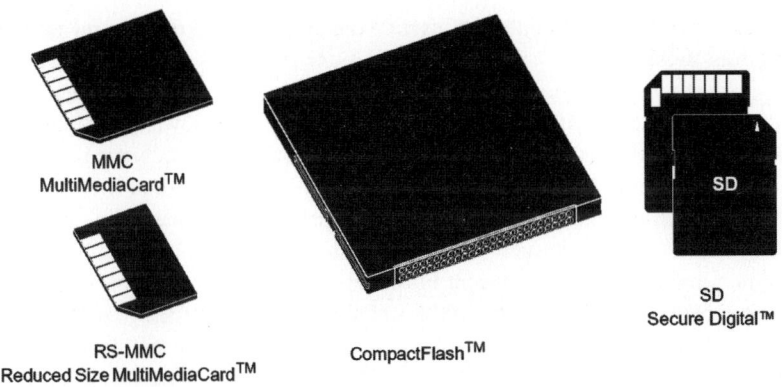

MMC
MultiMediaCard™

RS-MMC
Reduced Size MultiMediaCard™

CompactFlash™

SD

SD
Secure Digital™

Fig. 4.1 Different kinds of Flash cards

card operation and to understand how it fulfills the expectations of the surrounding worlds, it is useful to start with its internal content.

As suggested before, the memory device embedded in the Flash card is composed of one or more semiconductor memories, typically *Flash memories*, capable of being electronically written to and erased. These memories implement a low level protocol interface with the Flash card's external protocol. Furthermore, they need particular attention to guarantee their reliability, which is usual for any system that uses these kinds of memories.

For this reason, the Flash memory cannot normally interact with the external world directly,[2] but by means of an appropriate controller capable of adopting the role of interpreter between the external world and the Flash memories embedded in the Flash card. Figure 4.2 shows a simple scheme related to the internal content of a Flash card, in terms of the main blocks. The controller communicates with the host through the use of the specific Flash card protocol, translating the host's low level request for the Flash memory.

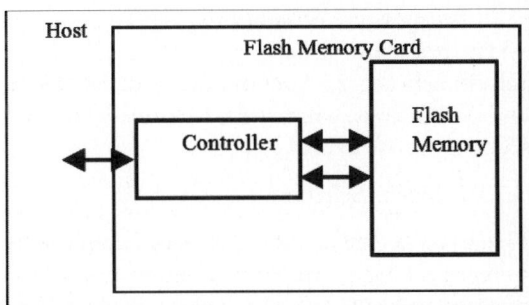

Fig. 4.2 A simplified Flash card scheme

[2] There are some examples of Flash cards without an internal controller belonging to a closed standard—that is, a situation in which both the Flash card standard and the host standard are managed by the same entity. This kind of approach comes with a lot of limitations.

The main operations the controller carries out in this scenario depend on the following conditions:

– the host communicates with the card through a logical address, while the memory has to be addressed from a physical point of view; this will be achieved through the use of *logical to physical remapping*;[3]
– the host can require the Flash card a level of performance that a single Flash memory cannot support; in such a case, the Flash card has to implement buffering techniques and multichip concurrent Flash memory access (if possible);
– the host depends on the reliability of the stored data, expecting data free of errors during the Flash card guarantee period; to achieve that result it will be essential to implement error correction, bad block management, and wear leveling techniques.

In the following paragraphs we will go in more detail about all the themes covered in this introduction; in particular, we will describe in detail the Flash card architecture, we will have an overview of the main standards specifications, and we will discuss the major Flash card applications.

4.2 Flash Card Description

4.2.1 Flash Card Architecture

From a logical perspective, a card is nothing but a nonvolatile memory device which communicates with a host through a certain number of interface signals determined by the protocol implemented. The protocol is the "language" by which a host and a card can communicate. Such communication is based upon two operations: saving data into the memory device (the write operation), and getting data from the device (the read operation)—both being executed without errors.

A first protocol classification can be done by taking into account the data bus width (parallel or serial protocol) and also the presence/absence of a sync signal (clock) between the host and the card. The MMC protocol, for example, is serial and synchronous, whereas the CF protocol is parallel and asynchronous.

4.2.1.1 Blocks Scheme

Figure 4.3 depicts an MMC architecture; such a structure can easily be extended to represent every type of card. From an architectural point of view, a Flash card is based upon the use of a dedicated controller ("card controller interface") and

---

[3] The *remapping* operation assigns a new physical address to a logical address, as will be described later.

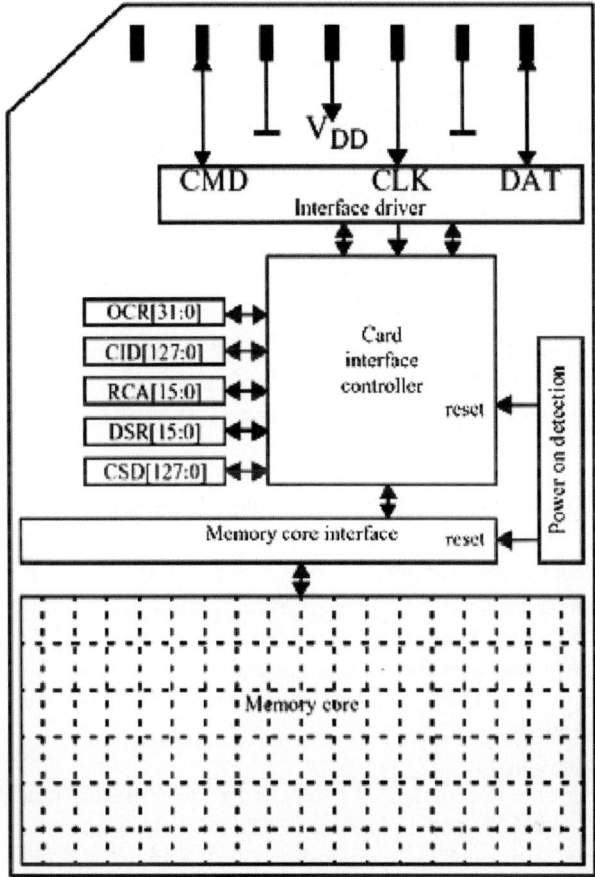

Fig. 4.3 MultiMediaCard™ architecture

a certain number of Flash memory devices setting up the nonvolatile storage area ("memory core" in Fig. 4.3).

The controller is interfaced on one side with the host and, on the other, with the Flash devices. Both controller and Flashes are soldered onto a common substrate. In Fig. 4.3, CMD, CLK, and DAT interface signals carry out the exchange of data and information between host and card. In particular, the CMD line is the channel by which the host sends commands to the card and the card, in turn, sends responses to the host (containing information about the correct command interpretation, or reporting any error). The DAT line is the channel by which data are exchanged.

Because it is a synchronous protocol, every host-card interaction is timed by a clock (the CLK line).

Before highlighting the architectural details of a card, it could be interesting to show why Flash memory is the most used technology in implementing the storage area on a card, and why a controller is needed to make the system work effectively.

Why Should I Put a Flash Memory on a Card: "NAND" Architecture As a Winning Solution

The use of Flash memory in memory cards is justified by the advantages that this technology offers in term of integration, power consumption, and endurance, compared to other types of nonvolatile memory.

EEPROM, for example, initially conceived and largely used to store the configuration data for devices like mobile phones and industrial control boards, is not suitable in applications for which a high storage capacity is required, although it is nonvolatile and reprogrammable. It is normally produced in sizes up to a few hundreds of Kbits.

Furthermore, since such devices are designed to save small amounts of rarely modified information, they do not tolerate frequent data updates (typical of mass storage applications) in terms of endurance and performance. Most of such devices have a simple serial interface.

In fact, Flash technology offers memory types which differ in terms of performance and reliability, being thus suitable and effective for different kinds of applications.

NOR-based Flash architecture, for example, serves well for storing code (software or firmware) which can be executed directly from memory itself. By contrast, NAND Flash is the state-of-the-art in mass storage devices.

NOR has a lower random access time for reading operations (in the order of a hundred ns, compared to ten µs for NAND), which makes it effective in reading and suitable for "code boot" operations.

NAND has a lower writing time (200 µs to program 512 bytes, in contrast with NOR, which needs the same time to program 32 bytes) and a lower erasing time (about 2 ms to erase 128KB in NAND and 1s to erase 32KB in NOR). This deters NOR usage in platforms like cards, which an application (host) accesses through continuous read/modify/write sequences, as if it were a floppy or a hard disk.

Because of this property, Flash NAND is used as the main storage device by the major memory card manufacturers (STMicroelectronics, Sandisk, and Samsung, to cite some of them). Since NAND technology is the most commonly used in the memory card market, it is appropriate to clarify in more detail which are the features that make it attractive.

First of all, the page dimension, both in reading and programming (512B or 2KB), matches, or is a multiple of, the minimum dimension of the logical unit defined by the ATA[4] standard: a sector (512B).

This is particularly attractive, because such cards are used in an environment supplied with an operating system and a file system, and the data transferred doesn't need any format adaptation to be read or programmed from or onto Flash. Performance is, in this way, maintained.

---

[4] ATA is the most commonly used protocol for hard disks. The logical information unit, both in transmission and reception, matches the physical unit in the storage device. This lets us simplify the mapping of the host's logical addresses into the Flash card's physical address space.

Furthermore, the size of a Flash erase block (16KB for 512B/page Flash type or 64KB for 2KB/page Flash type) has the same order of magnitude—if it is not perfectly equal to—a *cluster*,[5] which is the granularity used by a file system to split a file over a storage device, for example, a hard disk or a card.

Thus, rewriting already written clusters translates, internally onto the card, into erasing a whole number of Flash physical blocks.

Why Does the Card Need a Controller?

The job of the controller within the card is to translate a host interface protocol into a Flash interface protocol. Since, as already mentioned, different memory card standards implement different interfaces, the controller acts as a "bridge" between the Flash device and the host in such a way as to guarantee a reliable, consistent communication.

The fact that a controller is logically "in the middle" implies that it must implement all the functionalities foreseen by the two protocols.

The implementations of a controller can range from a finite state automaton to an 8-bit microprocessor, up to the most recent architectures based upon 32-bit controllers with dedicated peripherals.

The following considerations will help to understand how the increasing complexity of a host protocol on the one side, and the higher card performances required on the other side, affect the complexity of the controller.

First of all, the controller has to respond to commands generated by a host and translate them into reading or writing input commands for the Flash, within the standard timing stipulated by the protocol itself. The stricter such timings are (in high performance cards), the more it is necessary to find hardware solutions in order for critical functional blocks to work fast enough ("memory core interface" in Fig. 4.3).

The host accesses the Flash card as it would a hard disk; primarily, the application uses a Flash card to write and read files, but at the protocol level it interacts using the write and read commands with the cluster and sector data. In detail, the write operations can be repeated many times at the same logical addresses (for instance, during the file rewrite procedure), without any erase operation requested by the host side.

The controller has also the job of managing a set of operations called "garbage collection"[6] with the task of allocating space for the new data and rescuing space for the now obsolete data. The controller must hide from the host the slowness of these transactions, managing the associated tables in order to recognize, every instant, the Flash physical address of each logical sector sent by the host. All these operations normally form an algorithm called "file translation," with a computational complexity that requires the attention of a microprocessor; the algorithm is performed in program storage in a dedicated ROM memory that needs a RAM for the proper operands.

---

[5] This indicates the minimum amount of information writeable or erasable by a file system.

[6] The "garbage collection" will be explained in detail below.

corresponds to a higher card density that can be supported, and using the previous parallel strategy mentioned, results in better card performance.

Practically, we try to minimize the Flash memory dies used for the following reasons:

- often the card form factor is the main constraint for the number of Flash dies (primarily in the "reduced size");[8]
- both the controller cost and the complexity increase when the number of Flash dies increases, because the number of signals grows, as of course does the chip die size;
- the substrate cost or PCB cost is linked with the number of chips used (considering also the layer complexity and routing issues);
- the main directions to all card manufacturers regarding the density/price relation are:
  - to increase the Flash memory integration level considering technology changes (e.g., the current migration from 70nm towards 40nm),
  - to invest in advanced package technologies to allow the integration of many chip dies in the same space.

4.2.2 Controller Details

Here you'll find a description of an implementation of an MMC controller, and examples of the problems involved in the design of a memory card. As previously discussed, the controller of a *Flash card* is based on a microprocessor.

A basic constraint that we have to work with in designing this product is compatibility to the standards of the memory card. It is very important, therefore, to understand which are the functions that must be implemented by the hardware and which are the functions that may be implemented by the software. This activity could be considered as a system analysis in order to partition hardware and software blocks properly and it is very important, because errors in this phase may prevent the system from working well, even if it is possible to accept degraded performance. Choosing to locate a function in hardware gives the possibility of reaching high performance, but may reduce flexibility and the possibility of doing modifications if needed; on the other hand, implementation via a software function with high performance may have as a result not following the specifications.

The MMC, as was said before, has a serial synchronous protocol. The clock signal supplied by the host may have a frequency varying from 400 KHz up to 20 MHz. The host can manage the clock signal in different ways, not only continuously but also sending the clock signal only during the timing commands windows.

---

[8] Some Flash cards are originally designed with a large form factor, but, considering current interest in the mobile platform, it was decided to introduce a "reduced size" form factor; and at the moment there is a need to further decrease towards the "micro" form factor.

From this it is clear that the controller cannot be based only on a clock supplied by the host but must have an internal clock signal to guarantee the correct working of the operation independently of the host clock signal. This analysis requires some general consideration of the system. For example, if we have the target of very small consumption in standby mode, we have to introduce the "clock-gating" concept, and we need to switch off the internal oscillator and regulator (if present); as a consequence of this decision, we must face a delay in returning to ready after a command.

To effectively face the problem, it is possible to organize the system with one part that works off the internal clock and another part that works off the host clock; in this way, there is a part always active even when the internal circuitry is off.

4.2.2.1 General Architecture

Figure 4.5 shows the block diagram of the architecture implemented. The system has its core in the control unit using an 8-bit microprocessor; the other parts of the system can be thought of as peripherals of the microprocessor. The microprocessor is interfaced to the peripheral using registries; these can be registries whose programming provokes actions from part of the peripheral, or registries that allow knowing the state of the peripheral.

Beyond this there are *interrupts*, which are produced by peripherals when certain events take place. The data supplied from the host and those products from the microprocessor are moved through the device using DMA (direct memory access), which frees the microprocessor to directly move it, increasing performance.

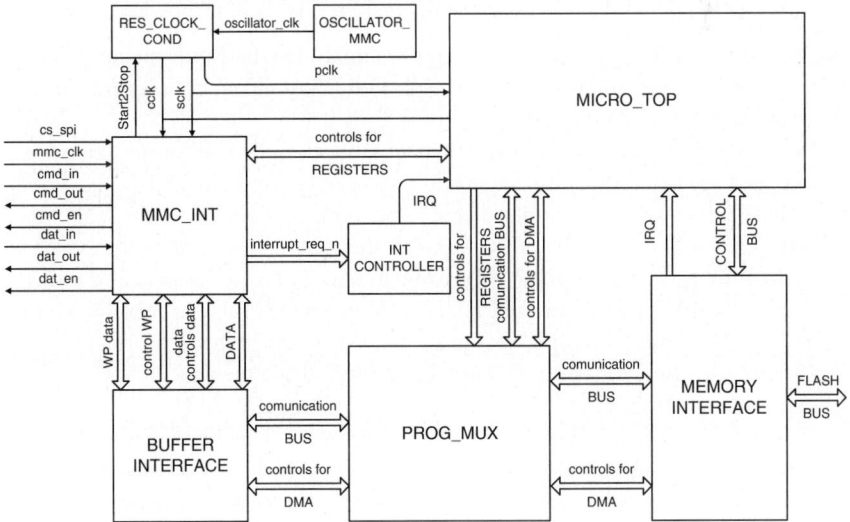

Fig. 4.5 Controller architecture

The MMC that we will describe in this text implements the protocol of communication with the host through more finite states machine that have the scope to interpret the commands from the host and to supply the relative answers. The management of the data is implemented through one programmable states machine. The ability to reprogram the algorithm is useful in this case because of the complexity of the protocol and the number of signals involved. This machine is a microprogrammed control unit; it is similar to a small microprocessor that executes an algorithm saved on logical gates. The entire MMC interface is moved by a clock supplied from the host, mainly because this guarantees answering commands in time, even if the internal clock is off and very slow to switch on.

The advantages of this choice are obvious from a consumption point of view, but have some complications in the interface management between peripheral and microprocessor. This type of organization introduces a synchronization problem: the information that goes from the interface to the microprocessor and vice versa is generated on different clock signals, so great attention must be paid to sampling that information in order to not lose information, with consequent wrong interpretations of commands and/or data. In order to avoid losing data, it is necessary that all information that changes clock domain must be processed by circuits that implement synchronization through an explicit "hand-shake," which assures the sender of the information that it has been received correctly.

The same problem is also present for the data. In order to synchronize the data, it is possible to use memory with double ports, which allows the use of one port with the host clock and the other with the internal clock, to guarantee synchronization (buffer unit).

The controller, as shown in Fig. 4.5, is composed of an interface to the Flash memory that implements the memory communication protocol. This interface must respect various constraints, mainly to adapt itself to the highest number of memories. It must allow interfacing the memory in all modes that are possible. Beyond that, it must be able to generate the control signals for the Flash memory, and it must be configurable, in order to adapt itself to the internal clock frequency and to the characteristics of the memory.

The memory interface must manage more devices in parallel at the same time. It must use the busy time of the memories during read, program, and erase to communicate with the other memories that are not working at that moment. This functionality, called "interleave," is very important because it raises the performance of the system: using this approach it is possible to exceed the performance of a single memory alone. This part of the controller is also in charge of the calculation of error correction of the data, which is necessary in order to guarantee high reliability of the system. This function must be done "on the fly," when the data are moved from the controller to the memory, so that it doesn't introduce any delay.

The controller is also composed of analog circuitries. These have to generate the right voltage supply for the logical core of the device and for the Flash memories. They also have to generate the clock signal used by the microprocessor and its peripherals. It is very important that those circuitries have a power consumption in stand-by null, or nearly so. The time used in escaping from the condition of low consumption must be small.

The management of the power supply is particularly delicate because the MMC must be able to adapt itself to the tension that it supplies when inserted into a host. It can be in either of two valid intervals: 1.65–1.95 V or 2.7–3.6 V. If the voltage is in the 3.3 volt range, the internal voltage must be regulated; if it is in the 1.8 volt range, the internal voltage must be supplied directly to the logic. All these operations must happen before the microprocessor executes the initialization of the logic.

During the stand-by to maintain the contained power consumption, it is a good idea to cut off the power supply to the NAND memories; this eliminates the memories' consumption during stand-by. In fact, even if it is very small, such consumption can become important when more memories are made part of the system.

Therefore the controller is able to switch the Flash memories on or off during the standard use of the Flash card. Also, the oscillator must respect hard constraints. In fact, it must supply a clock that remains very stable even in the presence of process, tension, and temperature variations, because from it the controller derives the signal with which it communicates with the Flash memory; the greater the error from the oscillator, the worse and less efficient is the communication with the memory.

The oscillator must be precise, but it cannot use external components from the chip. In fact, a big benefit would result from the ability to use external components as a reference; but it isn't possible, because this would increase the cost of the system.

4.2.2.2 MMC Interface

Figure 4.6 shows the architecture of the MMC interface. This interface uses the clock supplied from the host, and as mentioned in the previous paragraph, it implements the MMC protocol in two variants: standard and SPI protocol.

These two protocols are very similar; they differ mainly in the fact that MMC uses bidirectional lines: one for data and one for command; while SPI uses one line as input and one line as output on which to exchange command and data.

In the interface there is a module that allows choosing the behavior of devices from MMC to SPI (SPI_MUX). The interface is composed of one serialize/de-serialize, in order to transform the sequence of bits in the command and to transform the answers into sequence of bits. The commands are interpreted and, according to the type of commands, they can produce different actions: some commands can change the state of the Flash card, while others can program user data. This activity is done by the CUI that is substantially the heart of the interface.

The CUI is composed of two very complex finite state machines that interpret the commands and trace the current state of the Flash card, respectively. These machines must work with the information supplied by the commands; in order to do that they use an arithmetic logical unit that is able to make sums, removals, and comparisons; the results are used to determine the correct development of the Flash card. According to the commands received from the host the CUI switches on the microprocessor that generally is on only when the user requires a data transfer. The treatment of the user data, as already explained, is handled by a microprogrammed control unit (DAT_INT) that temporarily stores data in RAM before moving it into Flash.

memory physical organization: it creates a logical to physical memory mapping. The file system operates on the Flash card through the logical map. In this way, the Flash translation layer can emulate the sector rewriting operation (a classical feature of the hard disk), managing the logical map; the way it does so is by changing the physical location inside the Flash memory array which the *logical sector* point, and finally marking the old physical sector as invalid. The example shown implies a fundamental concept of FTL: the physical to logical sector remap[10] that consists in a one-to-one relation between host domain logical sectors and Flash domain physical sectors (Fig. 4.11).

The physical to logical remap which the FTL implements can be extended, in some firmware implementations, to a physical block too, even if the block is related to the Flash domain. In such a case, we consider the concept of *logical to physical remap of virtual block to physical block* (Fig. 4.12). During the remap operation, this can be needed to free memory invalid space, so that new free memory positions

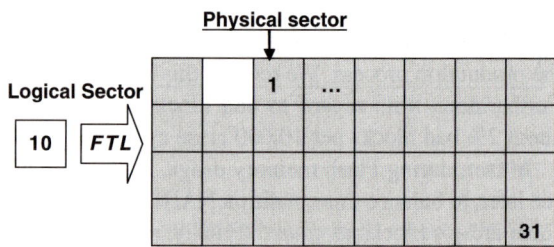

Fig. 4.11 Logical to physical remap

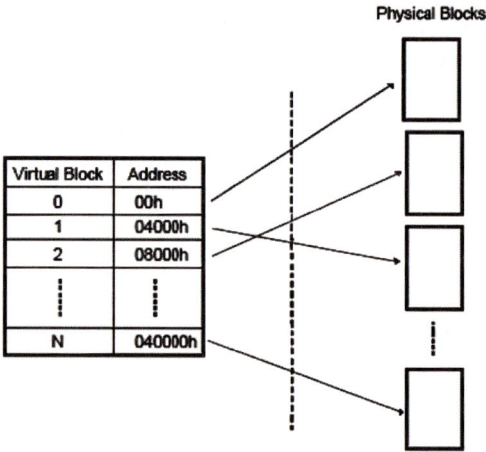

Fig. 4.12 Logical to physical blocks remap

---

[10] The host side addressable space can be represented as a logical map of the physical memory space; this way changing pointers to the physical area changes the logical map, whence the term remap.

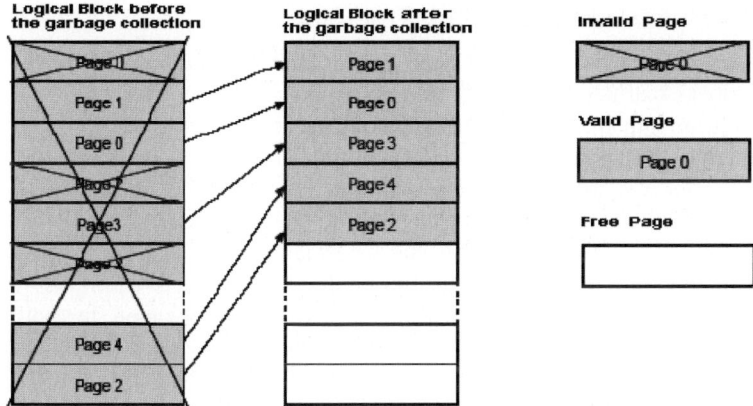

Fig. 4.13 Garbage collection

can host new physical sectors related to new writing operations. To this end, the FTL implements the garbage collection operation (Fig. 4.13), by means of which valid sectors, belonging to a well-defined Flash memory block, are moved into a Flash memory free block, while the old block, that now hosts only invalid sectors, can be erased.

Furthermore, some Flash card firmware implements *buffering* techniques, with the aim of minimizing the garbage collection operations, and consequently the erase operations, in the case of different sectors belonging to successive logical addresses, that is, at the same virtual blocks. To take into account the maximum number of erase-program cycles per physical block, the FTL can implement some garbage collection operations, with the aim of better spreading the block erase operations all over the physically available blocks; this technique is named *wear leveling*. It should be taken into account that a wear leveling management technique costs some system resources, in terms of microcontroller computation and in terms of Flash operations. For this reason, a good wear leveling algorithm is capable of implementing an acceptable trade-off between throughput and leveling efficiency.

4.2.3.3 Flash Memory Manager—Low Level Drivers

The high level host side Flash card access, by means of protocol commands, translates, after the suitable FTL elaboration, into low level Flash memory operations. These kinds of operations are executed by the microcontroller, by means of low level firmware drivers; otherwise this operation can be executed by certain more complex hardware peripherals. Nevertheless, such peripherals are driven by firmware that starts all the suitable Flash operations, checks the outcomes, to the extent of doing, in such cases, more parallel operations on different Flash memory chips, according to the maximum power consumption the host can supply.

4.2.3.4 Bad Block Management

Complete Flash memory bad block management, for both the *time zero* bad blocks and the bad blocks that develop during the Flash memory life, requires a suitable bad block firmware implementation. There exist different kinds of bad block management techniques, such as the *skip block method* or the *spare block reservation method*. The former is suitable to be integrated directly by the FTL, which, during the garbage collection operations, discards the invalid blocks in the physical allocation phase (remap) of a virtual block. The latter is normally implemented as a different firmware module—at the same level as Flash management drivers—completely avoiding the need for the FTL to deal with the bad blocks. In any case, for both the methods there should be available a recovery[11] function, whenever a new bad block develops during the erase or during a program operation. The recovery operation consists in saving valid sectors in a free memory block before it becomes invalid.

To conclude this section, it should be observed that in addition to firmware management there are other basic Flash card operations, such as the management of reserved commands (vendor's commands) for the purpose of customization, debugging, and testing, and analog block management, for the purpose of, for example, power saving system management during the stand-by condition. Finally, for some Flash card standards there exists a strong requirement related to the minimum performance that has to be assured to be compliant with the specification.[12] In such a case, one of the main firmware goals should be to fulfill this requirement.

4.3 The Standards on the Market

At the beginning of this chapter we discussed the different standards on the market; now we will set out a more detailed description. The main physical differences arise from the different form factors defined by the various standardization associations; but in addition, inside every specification there are different form factors, depending on the market addressed.

4.3.1 What Features Are Defined in the System Specification?

A standard is defined by a system specification document, which describes the package or container details, the communication protocol with the host application, the compliance or conformity rules (in order to guarantee the standardization alignment), the minimum performance required, etc.

---

[11] In this technical context, *recovery* means a set of operations to save valid pages.

[12] A performance parameter is the amount of read and write data per time unit (MB/s).

With the specification document, it is possible to design and manage a Flash-card device; but it is not as simple as it seems, because there are many different possibilities for implementing the behavior described, and every manufacturer can choose among several architectural strategies.

During the design process is difficult to cover all the compatibility issue generated by thousand of applications on the market but definitely the design successful it will depend also by the fulfillment of the application makers.

4.3.2 Why So Many Standards on the Market?

There is no real answer, but it seems that many standards will need to coexist for the next few years because different companies and standardization bodies will promote their own standards in order to win the market competition.

Usually, compared with other similar situations, the market is the real selector between the different form factors; but inside the Flash card world there is an anomaly, and this situation it seems will persist into the future.

There are many companies involved in the different standardizations currently promoted, and also in new standardization activities. Fortunately, the Flash card applications on the market can support a wide range of Flash card types.

There are about ten different Flash cards on the market; the differences are in the mechanical form factors, the electrical interfaces, and the targeted applications.

Every Flash card standard is managed by its own worldwide standardization body, which is in charge of the specification development, the patent policy, and sometimes the royalty rules.

4.3.3 MultiMediaCard™

This standard was introduced to the market after a collaboration between Siemens and SanDisk in 1997. Considering the mechanical form factor dimension, it is one of the better Flash card solutions ($32 \times 24 \times 1.4$ mm).

The applications addressed are: MP3 players, printers, digital cameras, and Smartphones; these, especially the last, are becoming the preferred market for this standard. For instance, the mobile phone based on the Symbian OS requires high density storage media in order to support its high capacity requirements, and the MultiMediaCard™ is the best compromise for these devices, especially in the Reduced Size MultiMediaCard™ at 1.8V power supply.

The MultiMediaCard™ has a synchronous interface using 7 pins for two serial protocols: the proprietary MultiMediaCard™ and the older SPI protocol (Fig. 4.14).

The latest specification revision created two new standards, MMCmobile and MMCplus, able to manage up to 13 pins (very similar to the Secure Digital™). The former will have the reduced size form factor and 1.8v power supply.

Fig. 4.14 MutiMediaCard™

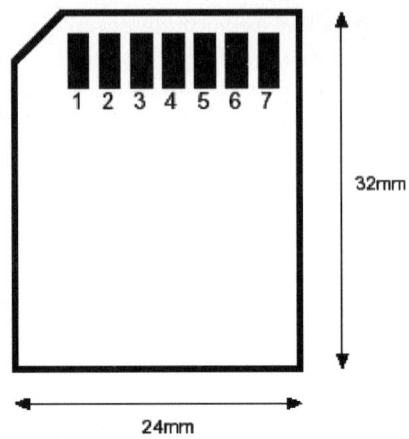

4.3.4 Secure Digital™ Card/miniSD™ Card/TransFlash™

The Secure Digital™ standard (Fig. 4.15) was presented during 2001 as a development of the previous MultiMediaCard™, although this new standard included some security features using the cryptographic method to protect the loaded content (CPRM) and parallel transmission on the DAT line. Also included were some mechanical differences in the thickness (2.1 mm) and an additional mechanical switch, as on the old floppy disk devices.

The functional protocols used by the Standard Secure Digital™ and the standard MultiMediaCard™ are really similar, as are also the speed performances. The actual development is going to move both cards in the same direction: dual voltage 1.8V/3V and parallelism.

The applications for this format are the same as those for the MultiMediaCard™ but with a larger diffusion in the market; the reduced mechanical form factor was also adopted in two other solutions: miniSD™ and TransFlash™ (Fig. 4.16). All the reduced form factors can be used as larger form factors using dedicated mechanical adapters.

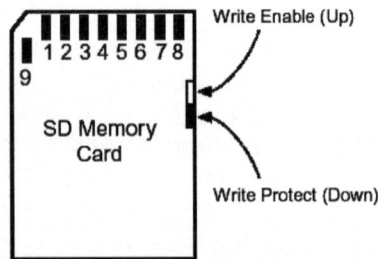

Fig. 4.15 Secure Digital™ card

Fig. 4.16 MiniSD™ and TransFlash™

4.3.5 CompactFlash™ Card

The CompactFlash™ standard was first introduced on the market in 1994 by San-Disk, based on the PC Card communication protocol. The mechanical dimensions are 36.4 mm × 42.8 mm; the thickness is 3.3 mm with the Type I form, and 5.0 mm with the Type II form.

The applications addressed are: professional digital cameras, printers, palmtop computers, and, especially, industrial devices. This last is becoming the preferred market because of the operating conditions required in industrial environments.

Three protocols are used: PCMCIA I/O Mode, PCMCIA Memory Mode and True IDE Mode, all based on a parallel transmission using 50-signal pins (Fig. 4.17).

Fig. 4.17 CompactFlash™ connector

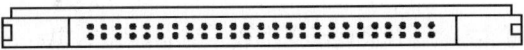

This standard keeps the advantage of a good ratio between performance and density compared to the others, but the large dimensions are in conflict with usage in MP3 players and other portable devices.

4.3.6 Memory Stick™

This was introduced in 1999 by Sony, the owner of this standard under a license and royalties agreement. There were different revisions due to limitations on the choice of design (e.g., lower memory capacity), but a stable version was reached with the model promoted by Sony/SanDisk called Memory Stick™ Pro in 2003. The card dimensions are 50 mm × 21.5 mm × 2.8 mm, and on the first models there are two switches: one to relieve the memory density limitation, and the second to prevent a cancellation risk. The connector is composed of 10 signals, and the internal digital architecture is similar to other Flash cards on the market, with one or more Flash memories inside bonded on a printed circuit, the whole assembled inside a plastic package (Fig. 4.18).

Memory Stick™ Pro performance is high but the high cost/byte and the low market share reduce the possibility for future growth in the market.

Fig. 4.18 Sony MemoryStick™

4.3.7 SmartMedia™/xD-Picture™ Card

The Solid State Floppy Disk Card (SSFDC) technology was developed by Toshiba and introduced to the market in 1996, after changing the name from SmartMedia™. Physically very similar to the old floppy disk, but with dimensions of 45 mm × 37 mm × 0.76 mm, the main difference with respect to the previous standardization shown is that there isn't any controller inside.

This type of package increases the integration capability of many memory modules, but all the software to manage the NAND memory properly are in charge the application side. For this reason the performance based on the physical data transfer is high, but the real logical data transfer is similar to the others' standard factors.

This element increased compatibility issues between applications, because the host could manage only some card densities and could not manage a higher density card. The standard is diversified with respect to the power voltage, identified by a notch on the left corner for 5 V and on the right corner for 3 V (Fig. 4.19).

The xD-Picture™ Card target applications are digital cameras. In the second half of 2002 a collaboration between Toshiba, Fuji, and Olympus produced the standard xD-Picture™ Card. The geometric dimensions are 20 mm × 25 mm × 1.7 mm, but the internal features are based on the previous specifications (Fig. 4.20).

4.3.8 USB Flash Drive

The USB Stick standard is not a Flash card but has many similar properties, like the dimensions, the portability, and the internal architecture (Fig. 4.21). It is mainly a storage device used as a floppy disk or as a CD. The protocol was based on the USB

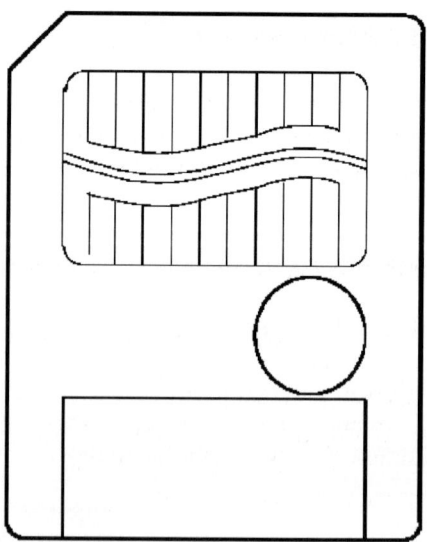

Fig. 4.19 Smart Media Card™

Fig. 4.20 xD Picture Card™

Fig. 4.21 USB Flash Drive

1.0 specification, but now the advanced USB 2.0 with better performance is very common. There isn't a common mechanical standard except for the USB connector itself. The body contains a controller, memories, and some passive components, and it is really similar to a marking pen.

4.4 Multimedia Applications

After this general description of the different Flash cards on the market, we are now going to describe the type of applications in which they are used and their related activities with other embedded memories.

4.4.1 What Is the Flash card Usage Model?

The major Flash card applications are usually mobile platforms like MP3 players, digital cameras, video cameras, palmtop computers, mobile phones, GPS systems, and digital car radios. There are also some peripherals used to share data by means of a PC adapter from the Flash card to a USB interface; not to mention also industrial applications like cash registers and telephone exchanges, using Flash cards as a replacement for older storage systems but mainly for their own easy interfacing capabilities.

In mobile applications the Flash card plays the role of main storage system, replacing the older magnetic and optical storage systems (audio cassette and compact disk). This allows the digital modification of data in real time (as in deleting a photo immediately after taking it, or replacing an MP3 file after playing it), thanks to the LCD display support.

All the applications start at boot without using the Flash card; but, if the Flashcard is connected, all the data contained inside are managed and ready for the user in only a second's time. Furthermore it is possible, in compliance with the application programming time, to make a "hot" extraction of the media without losing data, because in every mobile application there is an accessible casing like a drawer or a cleft.

We can divide the applications into two main groups: one in which the Flash cards are used to store data updated only rarely (palmtop and GPS systems), and the second used to store frequently updated data. In the first category the card contents are readable only, and the memory is called read-only memory (ROM), in which the software or data (such as road maps) are purchased directly with the Flash card or from an Internet secure transaction (using for instance the CPRM inside the Secure Digital™ card).

The second category is composed of devices requiring a frequent updating, like digital cameras, mobile phones, and MP3 players. The functional access to the card starts with an identification phase in which the host acquires the card properties (serial number, card density, available modes for writing and reading, and power consumption data).

After that the application is configured based on the card properties, but sometimes the application has a fixed configuration and an incompatibility between card and host could appear. In addition, only a few of the protocols are used by the application side in order to have greater compatibility with the cards on the market.

Most Flash card standardization documents don't contain the design rules for the application side; while most of the cards on the market have to be compliant with the standardization rules. Therefore it is important to verify the compliance standardization logo on the host platform. The data stored on the Flash card are arranged using a file allocation table (FAT) distinguished by card density, like FAT16 or FAT32. Be careful when using a Flash card greater than 2GB, because FAT16 can be used only on a card with less than 2GB density.

4.4.2 Cohabitation Between "Embedded" and Removable Memories

All the mentioned applications have some embedded[13] memories to store the code, or a little memory solution to store the data. For instance, digital cameras allow storing some photos on the internal memory if the removable card is not inserted. At the moment, the Flash card is used only as a data storage device, but there are some

---

[13] By "embedded" we are refering to the memory soldered onto the application board.

exploratory activities underway concerning bootability (the possibility of booting a PC using a CompactFlash^{TM} card in the PCMCIA slot).

References

1. MMCA Technical Committee (2004) MultiMediaCard system specification version 4.0. http://www.mmca.org
2. SD Group (MEI, SanDisk, Toshiba) (2000) SD memory card specification. Part 1. Physical layer specification, version 1.0. http://www.sdcard.org; http://www.sandisk.com/oem/transflash.asp
3. CompactFlash Association (2004) CF+ & CF Specification Rev. 2.1. http://www.compact-flash.org
4. http://www.memorystick.org
5. Toshiba Corporation (2000) SmartMedia^{TM} interface library—hardware edition; version 1.00. http://www.ssfdc.or.jp; http://www.xd-picture.com
6. http://www.usb.org
7. Moschini E (2003) Tecnologia di memoria: dalla teoria alla pratica. PC Professionale, 327–337
8. STMicroelectronics (2004) SMMxxxAF/DF/MultiMediaCard^{TM} target specification
9. STMicroelectronics (2004) SMMxxxAF/BF/SecureDigital^{TM} target specification
10. STMicroelectronics (2004) SMMxxxAF/DF/CompactFlash^{TM} target specification.
11. Niebel A (2004) Supply and demand Flash application-based five year forecast 2003–2008. Web-Feet Research, Inc.
12. Niebel A (2004) Flash card and removable storage forecast: 2002–2007. Web-Feet Research, Inc.

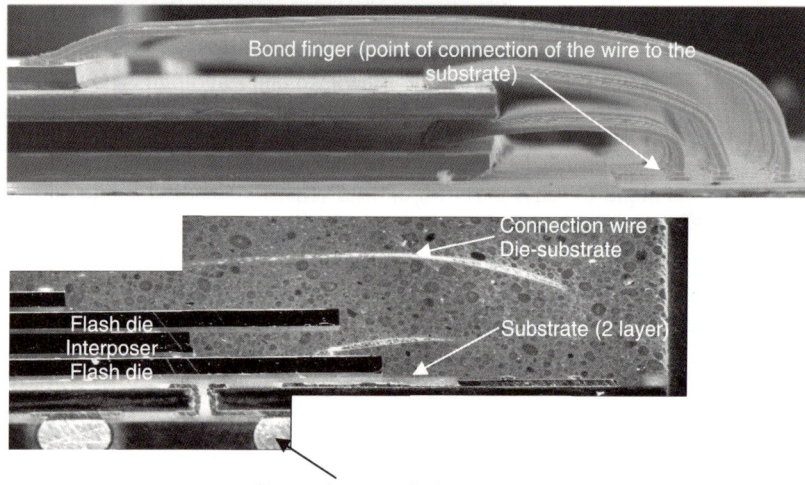

Fig. 5.7 Structure of the triple-stacked multichip

circuit between the wires of the lower die and the silicon above it, the upper dice are attached using adhesive tapes which cover the whole surface. Finally, the bonded devices are covered by an injected molding to protect the whole device.

5.2.2 Analysis of Preliminary Information

In this phase, a first compatibility analysis is performed on both the chosen devices and the chosen package, to make sure the final objective can be achieved.

The basic information includes:

- the specifications of the chosen devices,
- the sizes of the chosen devices,
- the pads' coordinates,
- the maximum available area for the multichip on the board of the final application,
- either a partial or an overall scheme of the package's external contact pattern ("ball-out" in case of a ball grid array).

For this specific example, a LFBGA88 package is considered: its size is 8mm × 10mm, and the external ball-out is shown in Fig. 5.8. The system block diagram is shown in Fig. 5.9, where it is evident that the data bus is shared by the three memories, and the address bus is shared as well, excluding A21 and A22, which are only used by the bigger Flash memories.

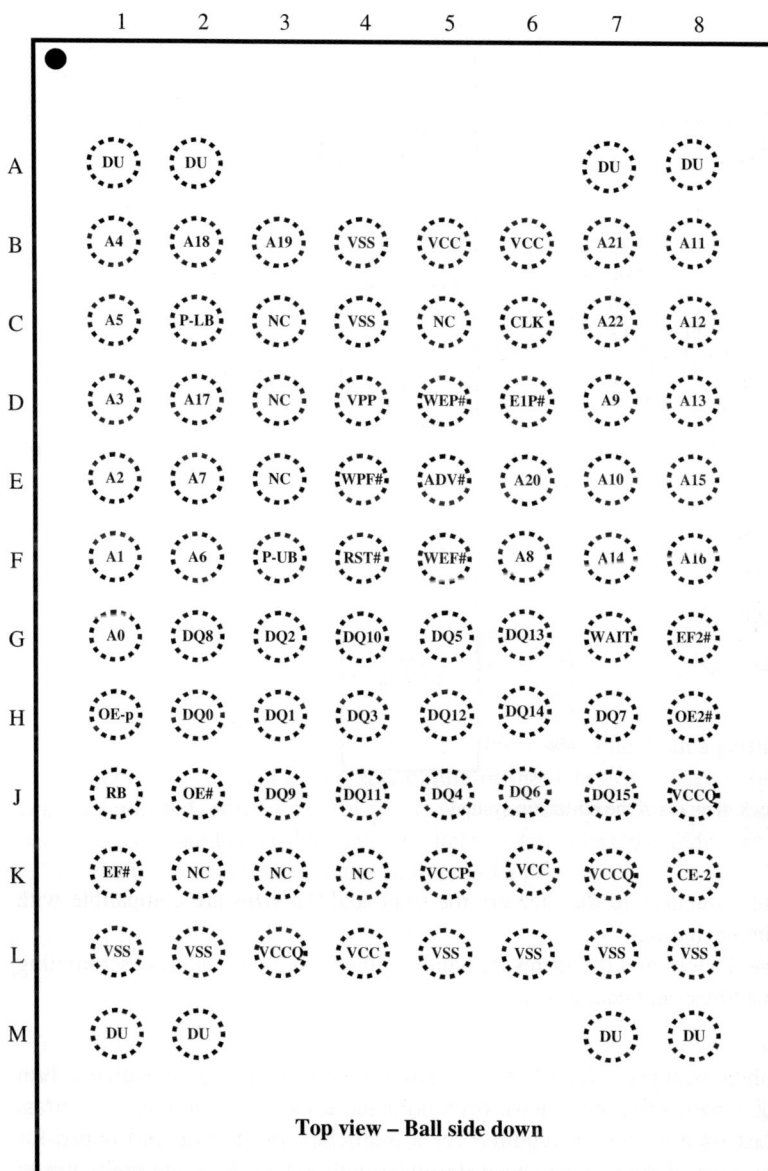

Fig. 5.8 Multichip "ball-out"

In order to have a routing of the copper traces of the substrate which is both feasible and the easiest possible, it is necessary that the highest number of bond fingers be shared; this means that two pads carrying either address or data in different devices should share the same signal. Such an objective can be achieved only in one of two cases:

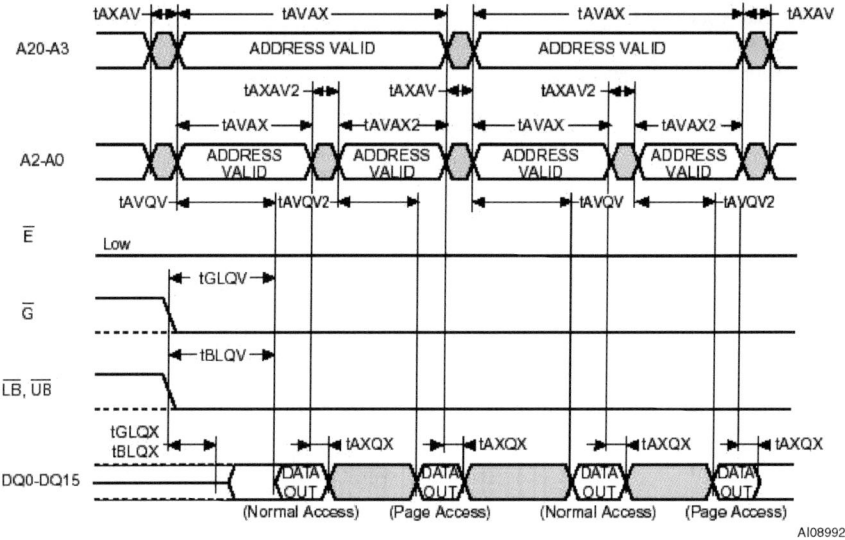

Fig. 5.11 Addresses A0, A1, and A2 must be identified in order to execute page mode read

As far as data scrambling[1] is concerned, it is necessary to recall the "x8" mode by dividing the 16-bit bus into two 8-bit buses by means of two control signals: "UB" (upper byte) and "LB" (lower byte).

There are two options:

- scambling inside the two bytes,
- scrambling inside the two bytes and their exchange.

In the latter case, used in the example, whoever generates the netlist (i.e., the files describing the connections inside the package) should connect the two signals of the volatile memory to the corresponding balls, inverting them (lower byte pad to the upper ball and vice versa); in this way, the chosen connection will be transparent to the user. Once the block diagram for the connections is set (as shown in Fig. 5.9), the system used to handle the feasibility flow is ready to automatically generate the corresponding netlists. The remaining topic is the intentional short circuit among the given external balls; every design tool short-circuits the balls bearing the same name, and the netlist must be coherent with this feature, which is fundamental to both the connection check and the signal integrity phases.

---

[1] Scrambling is the correspondence between the electrical address and the physical memory location inside the matrix.

5.2.3 Package Design

In order to ensure coherence between the connections inside the package and the package design, it is necessary to minimize human intervention in the flow of information provided for a multichip creation process. Therefore, all package design programs can automatically import a netlist in AIF format. This standard embeds all information needed by the multichip designer; every row of the file corresponds to a given connection, and it contains the following ordered data: the name of the external contact, the name of the pad to which the external contact must be connected, the pad coordinates, and the external contact coordinates. The file header contains information such as die size, pad size, and package size.

As already pointed out, the package designer has to import an AIF file for each die embedded in the target multichip. The software graphically shows the connections dictated by the AIF file, thus giving an immediate picture of the complexity of both the substrate routing and the gold wire connections between the pads and the substrate contacts (the bonding diagram). Despite the availability of routines for the automatic handling of the project, almost every multichip routing and bonding diagram must be generated manually. In fact, none of the tools available today are able to handle the usual design complexity, because it is often necessary to slightly violate the rules which were valid for previous designs. Such rules are not only dictated by the real physical limitations of the assembly equipment; they also result from previously assembled devices and from production yields.

The project outcome includes substrate routing, a bonding diagram for each device, package mechanical design and the cross-section.

5.2.4 Substrate

As already mentioned, the substrate is the printed circuit, typically composed of two, four, or more layers, which connects the bond fingers to the external contact balls by means of copper traces. Such start and ending points constitute the peculiarity of the substrate with respect to any other printed circuit. The solder mask (i.e., the protective resin) is opened on the innermost layer, where the first device lies, in order to make room for the bond finger sequence. On such fingers, as already said, the gold wires connecting the device pads are drawn. These pads are made of gold-laminated copper (typically $30\,\mu m$ of Cu + $10\,\mu m$ of Au) to ensure a better connectivity to the wires themselves.

On the outermost layer, the solder mask is opened to make room for the balls, which, in the case of a standard assembly, are composed of a Sn (63%) and Pb (37%) alloy. In order to allow a mechanically robust joint between the substrate and the ball, 3 different metal layers (copper, nickel, and gold) are deposited on the final pad of the copper trace, which improves welding (using a specific temperature pattern). The ball is usually attached by means of an electrolytic process; for this reason it is necessary to bring all the signals to the same voltage. All the traces are therefore short-circuited by connecting them to a single external trace, the plating bar. Traces will be separated at the end of the process (singulation).

5.2.5 Bonding Diagram

The bonding diagram is the schematic view of the layout of the devices and of their pads connected to the fingers underneath, as shown in Fig. 5.12.

For this reason, it is also the document used on the assembly line to give exact references to the person responsible for the correct connection of the devices. The more critical the bonding diagram, the more the assembly machines have to operate near their tolerance limit, thus sometimes lowering assembly yield. In the example case, lengths of the smallest device (32Mb PseudoSRAM) are the most critical issues.

5.2.6 Package Outline Drawing

It is the mechanical design which contains all the package dimensions. It is obviously related to the other elements of the design. For instance, the possibility of inserting a 4-layer substrate can be considered, or the possibility of narrowing the external solder mask opening can be implemented, thus increasing the portion of the ball exiting out of the substrate (stand-off). This trick is widely used when more degrees of freedom are provided to the substrate designer, thanks to the fact that the overall package thickness is not a major constraint. In fact, by narrowing the solder mask opening, the space between two balls increases, thus increasing the maximum number of copper routing traces that can be drawn between them.

5.2.7 Cross-Section

The package cross-section provides all the dimensions of the embedded devices, giving a precise idea of how exact the assembly technology must be. For instance, the backgrinding of silicon devices in the case of a quadruple-stacked multichip must be as thin as 50 μm!

In the example case, typical values are:

- 210 μm: substrate,
- 30 μm: resin to glue the first and the third devices,
- 105 μm: thickness of the first, second, and third devices,
- 75 μm: thickness of the silicon spacer used between the two Flash devices to allow wire connections to the lower die (twin-stacked),
- 30 μm: thickness of the tape used as glue between the first and second devices,
- 30 μm: maximum height (with respect to the die) of the wires of the first device (the "reverse bonding" technique),
- 60 μm: maximum height (with respect to the die) of the wires of the second and third devices (standard or forward bonding technique).

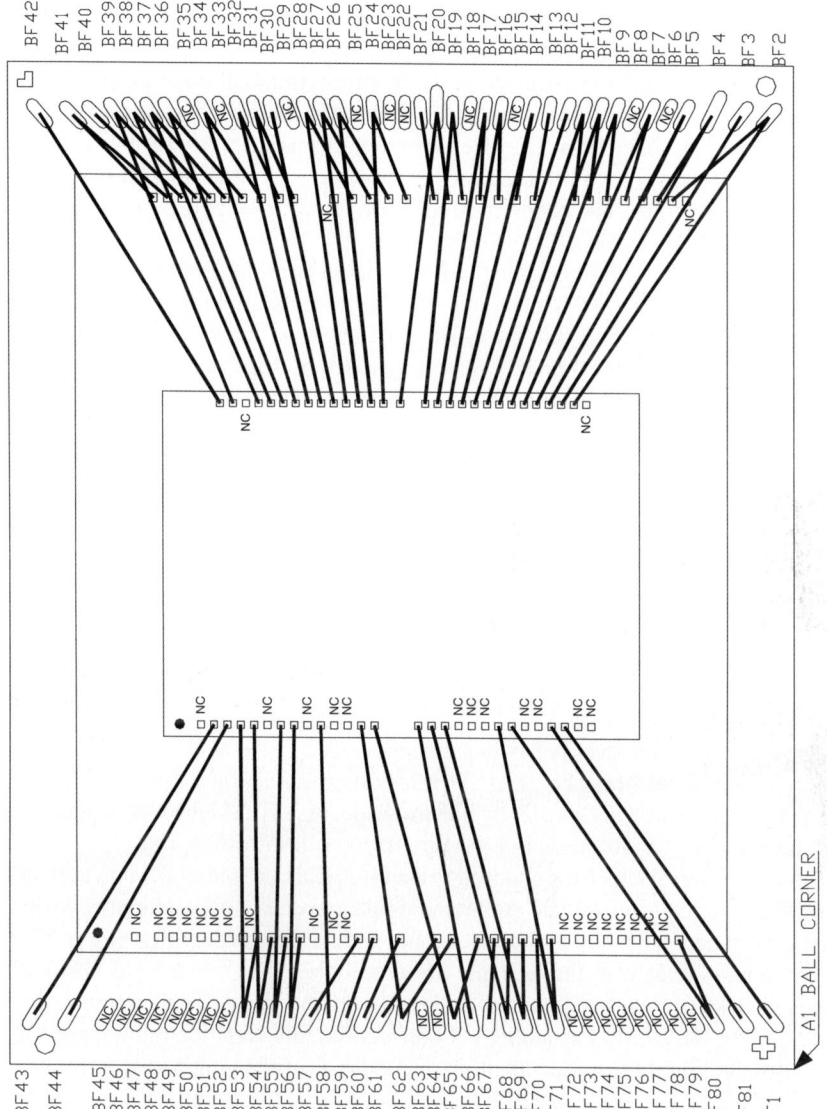

Fig. 5.12 Bonding diagram (Picture is not perfectly legible due to reduced dimensions, but it gives an idea of a typical bonding diagram)

5.2.8 Electrical and Mechanical Check

After the design is completed, it is necessary to verify its adherence to the required specifications. As in the generation phase of a new design, the check phase must be performed using a highly automated flow. In fact, any manual intervention can potentially generate an error. This principle is even more valid in the case of a multichip: for instance, there are almost 300 connections to be checked in the triple-stacked multichip example; even worse, such connections all overlap, since they share the same space! Needless to say, such a substrate cannot be manually verified; an automatic tool is mandatory. It is also important to guarantee that the two systems (project and checking) are not based on the same tools, in order to ensure that the whole flow is bug-free. The check is done in two steps: the logical phase and the physical phase.

In the logical phase, the checking software imports the AIF files (output AIF) generated by the design tool. The output AIF file contains not only the information from the input AIF file, but also the position of the bond fingers and the name of the ball (net) they lead to. For this reason, the output AIF files are also enough to generate the bonding diagram of each single device. Input AIF files are then compared with the output AIF files, and every discrepancy is reported as an error. The final assurance of the proper design of the substrate can be obtained in the latter phase, the physical check. The design tool also generates a file which describes the topology of each copper trace, in the way it will be handed over to the manufacturer. This file gets imported by the checking tool together with the AIF files, and a report file is generated, thus allowing the graphical preview of the substrate before it gets physically generated. Therefore it is also possible to check the path of each trace, also from a signal integrity point of view.

If the outcome of the check is positive, package parasitic extraction is performed: the result is re-elaborated and fed to the signal integrity tool, which can assess the real impact on the electrical behaviour of the device. A mechanical check is executed as well. The substrate, bonding diagram and cross-section for each device are imported and compared once again by means of specifically designed software tools which take into account both design and production rules.

Such a mechanical check (as opposed to the electrical one) must take into account the specific assembly site. In fact, every single rule is specific to the assembly device and its yield. There are many different mechanical rules, but in the following section only those rules related to the example are considered. The mechanical check of the triple-stacked multichip reported a violation of the design rule related to the wire length for the 32 Mb PseudoSRAM connection; this violation did not prevent volume production of the multichip, as explained in the following paragraph.

5.3 From Samples to Production

5.3.1 Feasibility and Manufacturability

Figure 5.13 schematically shows the assembly flow of the triple-stacked multichip from the example case. The first step is wafer backgrinding before it is cut. This operation cannot always be considered as part of the assembly flow, since in most cases it represents the last step of the wafer diffusion process. As a rule, wafer

Fig. 5.13 Assembly flow

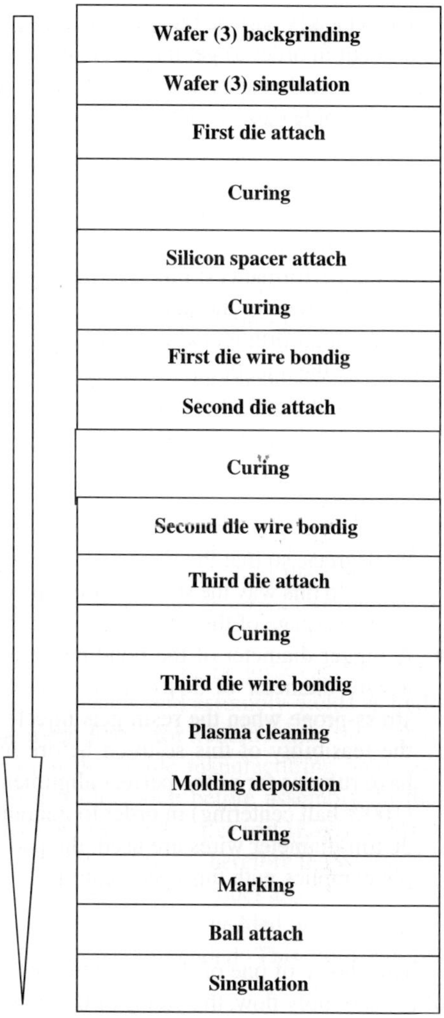

Wafer (3) backgrinding
Wafer (3) singulation
First die attach
Curing
Silicon spacer attach
Curing
First die wire bondig
Second die attach
Curing
Second dle wire bondig
Third die attach
Curing
Third die wire bondig
Plasma cleaning
Molding deposition
Curing
Marking
Ball attach
Singulation

backgrinding becomes part of the assembly flow only in the most critical cases, when the required thickness is less than 170 μm and the diffusion and assembly sites are not in the same plant; this is because shipping such a thin wafer is not recommended because of the high risk of breakage. After every die attach phase it is necessary to polymerize the glue further (including the glue on the tapes), which has been initially polymerized partially, in order to make it more workable. This operation is called "curing." Before the wire bonding phase it is necessary to clean any impurity by means of a high temperature baking process ("cleaning"). This operation is also repeated after the connection of all the dice, in order to improve protective resin adherence to the substrate. Devices are assembled on substrate strips

overcome if known good dice are used, where a higher test coverage is achieved "a priori" by different test flows on the wafers.

5.4 Evaluation of the Electrical Effects of the Package

The ever-growing demand for performance in wireless applications in the last few years has led to the design of devices whose operating frequency has grown considerably, thus reducing the switching time of the signals. If the operating frequency of the system is less than 10 MHz, the only aim when designing the substrate is to make sure all the connections are implemented. For higher frequencies, it becomes more difficult to guarantee proper device operation, because the switching time of the signals is so fast that all the contributions which are external to the devices embedded in the multichip (the package) can disturb the signal (sometimes endangering not only signal quality, but also its logical signal). In other words, differently from in the past, the package is no longer "transparent" for the signal, in particular for its higher frequency harmonics.

As a consequence, electromagnetic interferences inside a package, which are related to the electromagnetic fields generated by the electrical signals propagating on both the substrate traces and the bonding wires, can compromise the proper behaviour of a device. It is therefore mandatory to perform an analysis which can evaluate the effect of the package on the electrical signal and which is able to guarantee the correct logic interpretation of the signals on the package balls according to system specifications. This is the so-called *signal integrity analysis.*

One of the most important types of multichip is the mixed one, which is composed of different types of devices embedded in the same package to form a complex system. In a mixed multichip there are often buried connections, i.e., connections between pads which cannot be directly observed from the outside of the package. Such buried connections are not connected to the balls, therefore the electrical signal cannot be measured from outside. In these cases, the only way to ensure correct device functionality is to perform a signal integrity analysis, estimating the real electrical signal by means of simulations.

5.4.1 Substrate Interconnection Modelling

In order to understand the issues introduced by the package, it is necessary to describe how substrate connections can be modelled in a multichip, and what the limits for the validity of such modelling are. Substrate interconnection modelling is usually based on the classical model described in textbooks for the lines shown in Fig. 5.15a, which describes the equivalent circuit for an infinitesimal connection element dz.

In a substrate, the connection traces are separated by a dielectric, whose associated loss is usually negligible: parameter G'dz is therefore not taken into account in Fig. 5.15a.

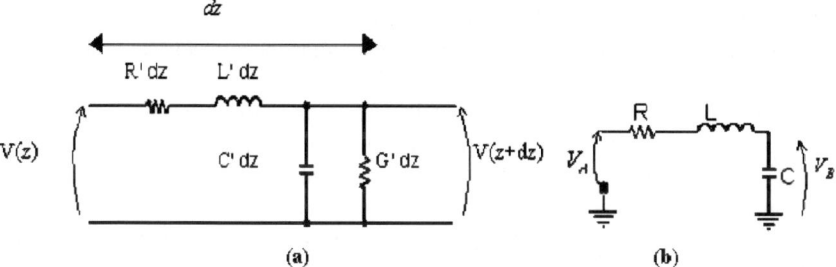

Fig. 5.15 (a) Classic modelling as described in textbooks for a line element dz; if line length is short enough (validity limits are explained later in the text), then the model shown in (b) is a good approximation for the behaviour of the line

In general, if f_{max} is the highest signal harmonic, the proposed model can be approximated by the model shown in Fig. 5.15b, assuming that the overall connection length is shorter than one tenth of the minimum wavelength, where:

$$\lambda_{min} = \frac{c}{f_{max} \cdot \sqrt{\varepsilon_r}},$$

c is the speed of light, and ε_r is the effective relative dielectric constant for the dielectric containing the connection. Both the spectrum and the band for a digital signal are elaborated in Sect. 5.8. In multichip manufacturing, the physical lengths of the connections are in the order of centimetres, and the maximum frequency for the spectrum component is lower than a GHz; therefore it is possible to estimate the signal on the output balls of the BGA and on the pads of the dice using a model similar to the one shown in Fig. 5.15b. Figure 5.16 shows the implementation of the proposed model to an input buffer (Fig. 5.16b) and to an output buffer (Fig. 5.16a).

The model shown in Fig. 5.16 has two limitations: first of all, the effect of the inductance on supply lines is not taken into account, which causes a modulation of the driving voltage of either the input or the output buffers of the device. Because of the variation of the current injected on supply lines, supply voltage can vary, and this kind of disturbance is usually referred to as *ground bounce*. In other terms, the source terminals of the MOSFET transistors of the buffers are biased to a potential which is not constant over time, and the current provided by the buffers depends on supply voltage. Therefore a negative feedback is present which, during signal switching, reduces the variation of the current provided by the buffer, thus increasing

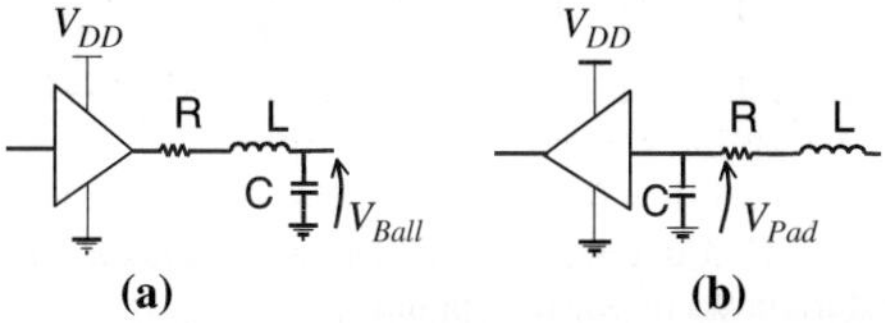

Fig. 5.16 Circuital model of substrate connections applied as a cascade to an output (**a**) and to an input (**b**) buffer

which meet both the capacitive and the inductive constraints. The substrate designer has the following geometrical degrees of freedom: the length and width of the connections and the distance between traces. On top of that, the following parameters can be modified: the maximum number of parallel traces, the number of supply lines used to break down inductive coupling between the different connections, etc. Even if substrate design is driven by a pre-layout analysis, it is mandatory to run a signal integrity analysis as well, once all the electrical interconnections have been drawn, in order to guarantee the proper behaviour of the device.

5.7 How to Perform a Signal Integrity Analysis

Signal integrity analysis, which is fundamental to substrate validation, consists of three phases. First of all, it is necessary to provide the circuital simulation with a model for each of the following elements: substrate connections, devices integrated in the package, loads driven by the buffers (for device outputs), and transmitters (for the inputs). In other words, three circuital blocks must be modelled: input and/or output buffers of the devices embedded in the multichip; the set of resistances, capacitances, and inductances representing substrate connections; and external transmitters and/or receivers driving (or driven by) the multichip. The features of the first circuital block (input and/or output buffers of the devices embedded in the multichip) are known to the developer and can be described using either a transistor-level model (if the exact architecture of the buffer is known) or the corresponding IBIS model, provided by the manufacturer of the die. The second circuital block, cascaded to the buffers, represents substrate connections (already described in previous paragraphs and derived by means of a magnetic field simulator able to extract package parasitics). The third block represents receivers, transmitters, and all those connections external to the multichip. As already stated in the previous paragraph, the multichip designer does not usually know the exact features of such a block. In the multichip datasheet, only the load that the buffers can drive (used to perform signal integrity analysis) is shown.

Once the circuital model is ready, behavioural simulation of the device is performed. It is essential to select the critical signals which are worth switching and probing, since it is virtually impossible to analyze all the possible combinations of signal switching, given the high number of signals of a multichip. The number of overall groups which it is possible to obtain from N signals is the sum of the combinations of N objects in k places, where k goes from 1 to N. Furthermore, the signals in each group might all toggle in the same way (logic level can toggle either $0 \rightarrow 1$ or $1 \rightarrow 0$) or not. In order to perform a complete analysis of the device, it is necessary to study all the switching combinations that may occur during device operation. Such an analysis requires huge computational resources, and today's equipment can cope with neither the time required nor the amount of memory. Furthermore the principle of overlap effect (i.e., estimating the effects of a set of switching events as the sum of the single ones) cannot be used, since there are negative feedbacks which modify system behaviour according to the amplitude of the signals under evaluation. The proper choice of both the groups and the related switching combinations to be analyzed is therefore of the utmost importance.

The third and last step of the signal integrity analysis is represented by the analysis of the simulation results. Simulated voltage signals are compared against appropriate masks to see whether they comply with required specifications. If some of the signals are out of specification, it is possible to modify the required connections in the substrate to fix the issue.

5.8 Digital Signal Spectrum

The spectrum of the signal switching in the system is one of the most important topics to take into account in the various phases of design, development, and analysis, since it is strictly related to the proper operation of the system itself.

When dealing with signal integrity, the aim of which is to verify that signal waveforms do not get altered by the filtering action of the package, it is mandatory to know the frequency pattern of the signals in order to establish good layout rules.

Once the intrinsic band of the signal(s) is known, substrate topology must guarantee that the poles introduced by the package ideally act at frequencies which are higher than the highest harmonic components of the signal. In reality, a reasonable goal is to guarantee that most of the energy content of the signal falls within the passing band of the substrate; all those harmonic components which get filtered out by the substrate must be negligible enough not to compromise the information contained in the signal. In the following discussion regarding spectrum estimation, the timing signal (a trapezoidal-shaped pulse train) is considered, and the rise time of the step is finite.

From these assumptions, it is possible to derive some results whose validity is general.

As reported in the literature, the spectrum of an ideal pulse train with duty cycle 50% and with frequency $f = 1/T$ is composed of odd harmonics with respect to the fundamental one; the amplitude of these harmonics is enveloped on the sinc(x) function. In cases in which a 50% duty cycle is not guaranteed, even harmonics appear in the spectrum too. The unilateral spectrum for the ideal pulse train is shown in Fig. 5.19, and it is composed of a pulse train whose amplitude is c_n.

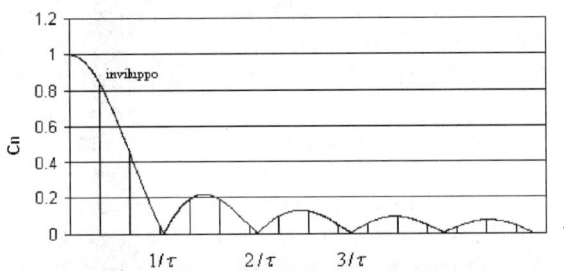

Fig. 5.19 Unilateral spectrum of the pulse train

Fig. 5.20 Trapezoidal-shaped signal

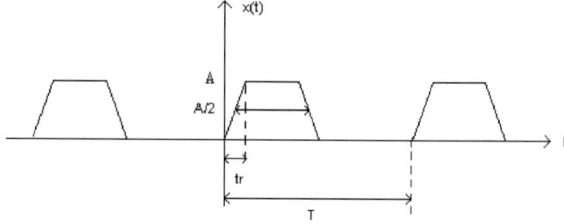

C_n quantity represents the normalized modulus of the nth spectral components of the signal; if the signal becomes the one shown in Fig. 5.20, it can be proved that the coefficients of the Fourier series expansion can be expressed by (5.1).

$$|c_n| = \begin{cases} A\dfrac{\tau}{T} \left|\operatorname{sin c}\left(\dfrac{n\pi\tau}{T}\right)\right| \left|\operatorname{sin c}\left(\dfrac{n\pi t_r}{T}\right)\right| & \text{per} \quad n \neq 0 \\ A\dfrac{\tau}{T} & \text{per} \quad n = 0 \end{cases}. \tag{5.1}$$

A good method to evaluate the band at 3 dB for this signal is to use masks. Since the following inequality (5.2) is always true:

$$\left|\frac{\sin x}{x}\right| \leq \begin{cases} 1 & \text{small values of x} \\ \frac{1}{|x|} & \text{big values of x} \end{cases}, \tag{5.2}$$

it is possible to build a mask for the sinc(x) function by means of 2 asymptotes as shown in Fig. 5.21.

The former asymptote equals 1 and has a slope of 0 dB/decade on a logarithmic plane; the latter decreases like $1/x$ and therefore it has a slope of -20 dB/decade. By replacing $f = n/T$, the envelope function (5.3) can be obtained.

$$envelope = A\frac{\tau}{T} \left|\operatorname{sinc}(\pi\tau f)\right| \left|\operatorname{sinc}(\pi t_r f)\right|. \tag{5.3}$$

The mask under analysis is the logarithmic diagram shown in Fig. 5.22.

Two poles can be recognized, one at $\frac{1}{\pi f \tau}$ and another at $\frac{1}{\pi f t_r}$. It is evident from this diagram that if rise time t_r decreases, then the second pole moves toward the higher frequencies, and therefore the amplitudes of the harmonic components included in the mask increase. On the other hand, if rise time increases, then the second pole

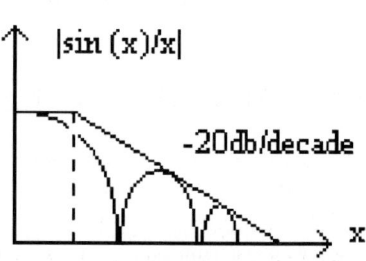

Fig. 5.21 Mask for the sinc(x) function

Fig. 5.22 Mask for the
spectrum of the trapezoidal
signal

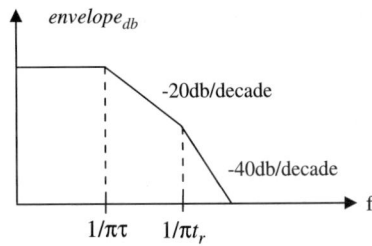

Fig. 5.23 Fourier transform
modulus for a unit-amplitude,
unit-rise time step function

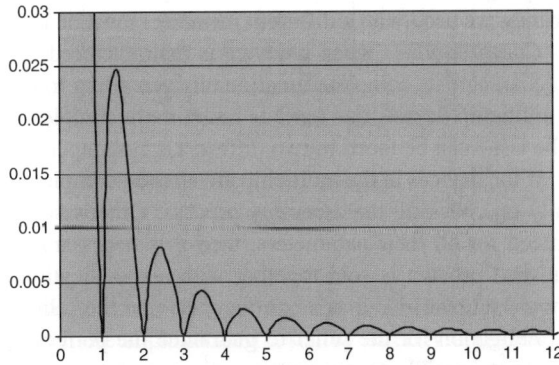

moves toward the lower frequencies and the amplitudes of the higher order harmonics of the trapezoidal signal decrease. Therefore the spectral content at high frequency for a periodic trapezoidal signal mainly depends on both the rise and fall time of the pulse. Non-ideal step function (i.e., finite rise time) does not have a Fourier transform in the strict sense, since it is a finite-power signal; but it is possible to give a representation of the Fourier transform by means of the theory of distributions.

As expected, the dominant component for this signal is the continuous component; other components are due to the transience of the signal itself. Therefore a Dirac pulse in the origin, with an amplitude of A/2 (where A is the regime amplitude of the step), can be expected as a representation of the modulus of the transform, plus some side lobes.

It is quite easy to prove that the envelope of these lobes rapidly decreases with the square of the frequency. Figure 5.23 represents the modulus of the Fourier transform for a unit-amplitude, unit-rise time step function, around the origin.

5.9 Multichip Testing: Observability, Controllability, Reliability

The world of multichip represents a new frontier for system integration, but it has to comply with the rules and constraints of large scale manufacturing.

Observability: every function inside the multichip must be accessible. Let's assume that the multichip system comprises a digital processor and a memory. From

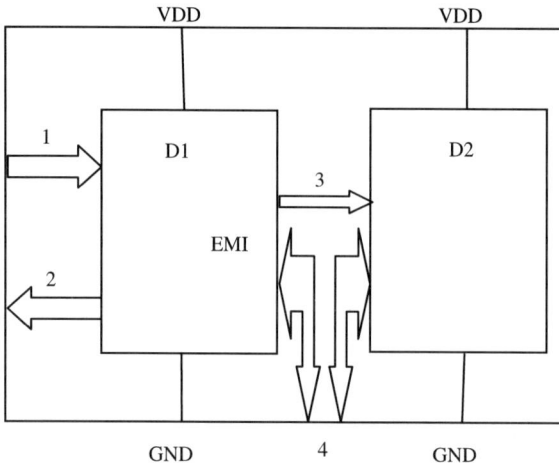

1) External Inputs

2) External Outputs

3) Memory Addresses and some Memory Commands

4) Outputs and Commands for the Direct Memory Access diretto alla memoria

Fig. 5.25 A possible solution

either separated (and therefore they are externally connected in order to rebuild their functionality) or connected, depending on whether or not it is possible to set the signals in high impedance in order to drive them from outside (because of an electrical incompatibility during the test, e.g., voltages which are too high for the driver of D1).

As mentioned previously, one way to optimize a multichip project consists in developing the project itself embedding all the components, i.e., both the devices and the substrate are an integral part of the design. Such a solution could imply both dedicated buffers, in order to ensure the integrity of all the signals exchanged between the devices, and specifically designed test structures. A potential evolution is either the integration (in case they are not present yet) or the reuse (in case they are already present) of specific self-test units, thus loading test programs which are dedicated to the system under test (embedded microcontrollers).

Another evolution could be represented by the introduction of both A/D and D/A converters to change analog signals into digital ones, thus minimizing the number of signals to be brought outside to deal with reliability issues. In the future, it might be necessary to devise methods of physical inspection for buried and nonfunctionally accessible signals (e.g., test-points which are covered by the solder mask), and to extend boundary-scan logic to every device in order to perform a connection check. Such an approach would greatly simplify connection checks in case of shared buses and buried connections. In the latter case, the check is performed using functional access, but the testing time required is often too high.

5.10 Vertical Integration

In the last decade, a huge effort, both technological and economical, has been made towards the miniaturization of electronic components. This activity has led on the one hand to the reduction of the minimum dimensions for the lithographic process, and on the other hand to the integration of system functional blocks into a single package. The latter topic can be divided further into two development paths: one the integration of such parts into a single silicon chip (thus leading to the so-called System-on-Chip, SoC); the other the integration of such parts, diffused on different silicon chips, into a single package (the so-called System-in-Package, SiP). Development of System-on-Chip started before that of System-in-Package, in the mistaken belief that it would have been possible to integrate all the functions in a single silicon chip very quickly, and with no penalty in terms of cost, performance, or time-to-market. In reality, the integration of different technologies always involves huge technical and economical efforts, which are not always justified by the performance of the final system; furthermore, project flexibility is no longer achievable, since a small change in a block leads to a rework of the whole system. For these reasons, System-on-Chip is now relegated to market niches where absolute performance is the driving factor. Development of System-in-Package is a variation of System-on-Chip: system miniaturization is still the main goal because of its advantages, but the single components are still designed separately and then integrated at the assembly stage. In this way, most of the disadvantages of System-on-Chip are no longer present (the development and optimization of both technologies and design, the time and cost of development, flexibility), while similar results in terms of miniaturization can be achieved.

SiP, in particular one in which the different dice composing the system are piled vertically inside the same package, is currently manufactured and used in a wide variety of highly integrated devices, like mobile phones and portable applications. A relatively easy way to get such a vertical system consists in stacking the dice inside the package and connecting them (using the wire bonding technique) to a common substrate, in which the external signals are then drawn, as shown in Fig. 5.26).

In any case, future applications will require an even higher set of functions from the system and, obviously, a smaller physical size. As an example, Fig. 5.27 shows the overall schematic of a mobile phone and those parts which, today, can be either inserted in a SiP or integrated in a SoC. All the R&D facilities in the mobile phone business are working toward the development of highly-integrated, three-dimensional (3D) systems.

Fig. 5.26 SiP composed of single dice vertically piled and assembled in the same package

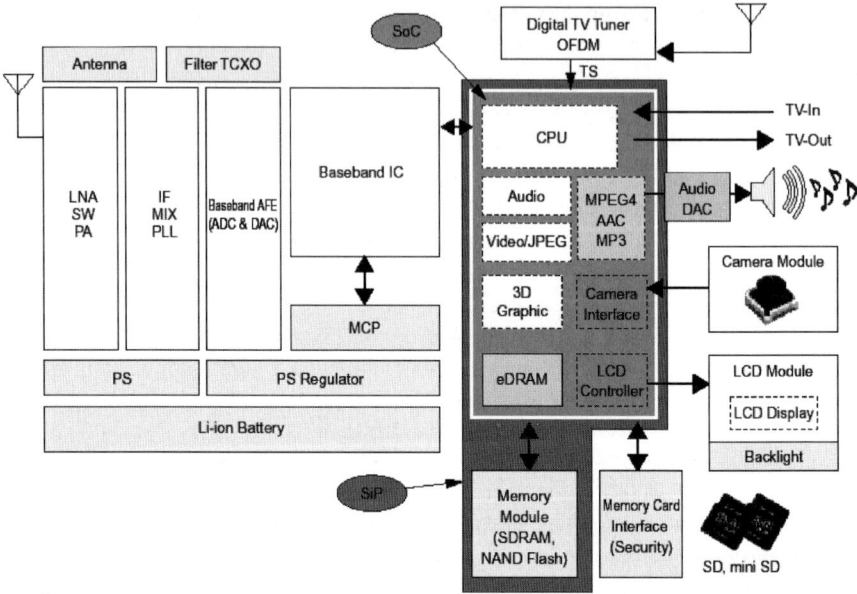

Fig. 5.27 Overall schematic of a mobile phone

It is not possible to cover in this book all the ongoing developments in detail. Nevertheless, some significant examples of solutions will be discussed.

5.10.1 Package Stacking

Cost and reliability are the main issues related to the SiP (a pile of dice inside the same package) from an industrial and manufacturing point of view. Indeed, as described in this book, both electrical and functional features are verified, in the normal production flow of an electronic device, before the inclusion of said device inside the package. On the other hand, only the main functionality will be verified after the assembly in the final package. This is because the test process performed on the device before packaging requires less expensive equipment. In some extreme cases, some functions cannot even be verified after assembly.

In the field of wireless applications, a system composed of a processor and a memory (both volatile and nonvolatile) is one of the most common configurations. In such a combination, in which both high value-added components (the processor) and low-margin commodities (the memory) are associated, some of the internal connections are often not accessible from the outside. As a consequence, once the system is assembled it is not possible to access all the functions of a single part. This condition implies a great economic loss, since all the issues introduced on a single part by the system assembly cannot be verified systematically.

Therefore the failure of a part, for instance the commodity, will lead to a failure of the entire system. In order to solve this issue, new technological approaches, alternatives to die stacking, are emerging. In these cases, the system still requires the stacking of component parts, but each part (or smaller set of parts) is assembled in independent packages, thus leading to a vertical stacking of packages. In this way every component can be electrically tested before becoming an integral part of the system; moreover, the impact of the assembly process on the single part can be verified.

The stacking of packages technique can be implemented in two ways: Package-on-Package (PoP), shown in Fig. 5.28, in which the dice have the same planar size and the connections are achieved by means of a contact ring on the perimeter of the package; and Package-in-Package (PiP), shown in Fig. 5.29, where the single, packaged, tested components are embedded together in another package.

As shown in Fig. 5.28, the Package-on-Package technique does not require the adoption of innovative technological steps; standard package technology is used. The peculiarity lies in the possibility of performing complete electrical analysis of every independent die, accessing all the possible functions and connections. When compared to the stacking of parts inside the same package, a larger planar footprint can be seen, because the connection among different packages is realized using the contact ring running along the perimeter. On the other hand, the contacts between the lowermost component and the board on which the system is mounted can be placed on the whole lower area.

When a PoP structure is used, the reliability test should focus on mechanical board stress (i.e., the board's endurance with respect to mechanical and thermal strain). PiP and PoP differ in several ways:

- In PiP, the connection between the parts is realized by means of wire bonding, which is a well established technique in packaging.
- In PoP, the connection between the parts is realized by means of board mounting techniques, and such process steps are known in themselves, but have never been used in this context.

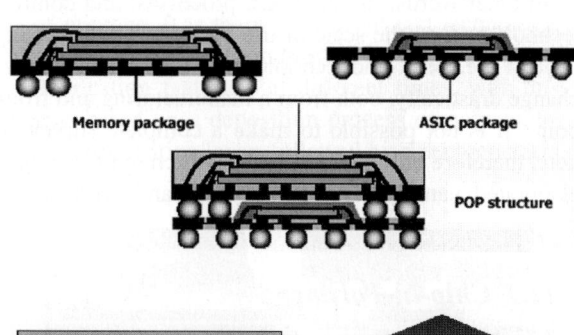

Fig. 5.28 Example of
Package-on-Package

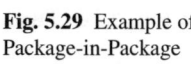

Fig. 5.29 Example of
Package-in-Package

definition through the substrate which allows reaching the device underneath. After an isolation process of the contacts with respect to the substrate, by means of a series of oxide depositions based on different techniques, the contacts themselves are filled with several metal layers (barrier layers like titanium nitride, tungsten or copper). Once the lower end of the substrate is reached, the newly-formed metal contacts are connected to the device underneath through some areas which are covered by several metal layers as well. The connection of the two devices between the metal layers is achieved by means of a welding process which generates a so-called intermetal interface layer (in this case Cu_6Sn_5), which acts both as electrical contact and mechanical grip.

Such a technological approach will allow a very high level of vertical integration, even though all the reliability and manufacturability issues already discussed will require a deep investigation. In conclusion, coming applications will require more and more complex and sophisticated system integration; any solution under investigation will turn out to be applicable not only if its feasibility and reliability are proven, but also only if it is viable from both a cost and an investment point of view.

References

1. R. Miracki, et al., "Rapid prototyping of multichip modules," Proc. IEEE-MCMC Multichip Module Conf., (Santa cruz, CA), March 1992.
2. H. B. Bakoglu, Circuits, interconnections, and packaging for VLSI, New York, Addison Wesley, 1990.
3. L. L. Moresco, "Electronic system packaging: the search for manufacturing the optimum in a sea of constraints," IEEE Trans CHMT, Vol. CHMT-13, Sept. 1990.
4. C. E. Bancroft, "Design for assembly and manufacture" Electronics Materials Handbook, Vol. 1 Packaging, Materials Park OH, ASM International, 1990.
5. L. H. Ng, Multichip modules cost: the volume-yield relationship, Surface Mount Technology, March 1991.
6. L. Higgins III, "Perspective on multichip modules: substrates alternatives," Proc. IEEE-MCMC Multichip Module Conf., (SantaCruz, CA).
7. G. Geschwind and R. M. Clary, "Multichip modules: an overview" PC FAB, Nov. 1990.
8. T. J. Moravec, et al., "Multichip modules for today VLSI circuits," Elect. Packaging and Prod., Nov. 1990.
9. K. P. Shambrook, E. C. Shi, and J. Isaac, Bringing MCM technology to board designers Printed Circuit Design, Vol. 7, no. 10. Oct. 1990.
10. R. C. Marrs and G. Olachea, "BGA's MCMs," Advanced Packaging, Sept./Oct. 1994.
11. B. Young, Digital signal integrity, Prentice Hall, 2000.
12. D. Crowford et al., "Applying physics based analysis to the design and verification of high frequency integrated circuits," Euro Design Conference 2004.
13. Ansoft Q3D Extractor.
14. M. Shoji, High digital circuits, Addison Wesley, 1996.
15. C. R. Paul, Compatibilità elettromagnetica, Hoepli, 1999.
16. H. B. Bakoglu, Circuits interconnections and packaging for VLSI, Addison Wesley, 1990.
17. C. Prati, Teoria dei segnali, CUSL 1997.
18. W. D. Brown, Advanced electronic packaging, IEEE Press, 1999.

19. G. P. Vanalli et al., "Fully testable double stacked logic plus flash with embedded connections," IMAPS, Pori, Finland, 2004.
20. Internation technology roadmap for semiconductor, assembly and packaging http://public. itrs.net/
21. G. Campardo, R. Micheloni, D. Novosel VLSI-design of non-volatile memories, Springer Series in Advanced Microelectronics, 2005.

Chapter 6
Software Management of Flash Memories

A. Conci, G. Porta, M. Pesare, F. Petruzziello, G. Ferrari, G. Russo
and G. D'Eliseo

6.1 Introduction

In recent years, Flash memories have been consolidating their important role in wireless and automotive systems, in contrast to the previous expensive and low density EEPROM memories and the too inflexible ROM (used only to execute one-time-programmable code).

There are mainly two kinds of nonvolatile information usually stored in a Flash memory: system code, which is the binary image of the operating system and the applications to be run on the device; and data, representing both system parameters and user information.

According to their memory array organization, Flash memories are categorized into different types, with correspondingly different peculiarities. The most diffused types of Flash memories are the so-called NOR type and the NAND type. The main parameters to be considered in a comparison between NOR and NAND Flash memories are throughput in sustained conditions, random access time, and erase time, in addition to the different interface philosophy (memory mapped for NOR Flash

A. Conci
Numonyx, Agrate Brianza, Italy

G. Porta
STMicroelectronics, Grasbrunn, Germany

M. Pesare
Qimonda, Munich, Germany

F. Petruzziello
Qimonda, Munich, Germany

G. Ferrari
Numonyx, Naples, Italy

G. Russo
Numonyx, Naples, Italy

G. D'Eliseo
Numonyx, Naples, Italy

R. Micheloni et al. (eds.), *Memories in Wireless Systems*,
© Springer-Verlag Berlin Heidelberg 2008

and I/O mapped for NAND Flash). These characteristics differentiate the typical application fields of the two categories.

With a very low access time (on the order of 10 nanoseconds) and the memory mapped interface, NOR memories allow direct execution of system code without usage of volatile memory (e.g., in common PCs the code stored in hard disks, solid state disks, storage media, or nonvolatile cache memories is first uploaded into the central RAM and then executed, because these media do not allow direct execution).

NAND memories, on the other hand, do not allow direct fetching of instructions and have a slow random read speed; these devices are intended to be used for high speed transfer of big amounts of data and can only be used to store binary code and then download it to the main system memory. This model of code execution, similar to what happens on common PCs, is called *Store and Download* and consists in loading in the main RAM either the whole binary image to be executed (during the bootstrap) or the pages of code wanted according to the system status and needs (*Page Demanding*).

In terms of code storage, the disadvantage of NAND Flash is its usage of RAM, which means that the system needs a bigger memory for the code image (or else a smaller part of memory is available for dynamic variables, buffers, caches, etc., if the same amount of RAM is mounted). Typically, more RAM also means more power consumption for the *refresh* mechanism.

NOR memories, on the other hand, allow the so-called *Execute-In-Place* (*XIP*) mechanism, which means code is directly executed from the nonvolatile memory, limiting the RAM size.

For purposes of data storage, on the contrary, NAND memories have a much better throughput, in particular for writing, due to the higher program speed and the much lower erase time. In addition, NAND memories have a better integration capability due to the array organization (space is saved because there is no need for the drain contacts needed for the NOR geometry), and so typical NAND devices are smaller than NOR ones.

In conclusion, NAND memories represent the best choice for data storage, while NOR memories allow XIP and, therefore, require less use of RAM for code execution. The definition of the entire system architecture and the kind and size of memory to install on it is a consequence of a lot of parameters, including performance, cost, reliability, reuse of legacy IPs, etc.

6.2 Software for Data Management on Flash Memories

The storage of a single byte (8 bits) or a string of words (16 bits each) is only the last step of a much more complicated process. The entire system has to be able, for example, to reconstruct a file starting from a set of fragments, or it has to interpret a bit sequence as a part of real data or as a control structure or, at the end, to guarantee the robustness of the storage process and data integrity.

Stored data are a logical entity while their corresponding storage fragments on Flash are a physical sequence of bits; this logical to physical association is the main task of the software for data storage on Flash memories.

This software has to take into account the different peculiarities of NOR and NAND devices. In addition, application requirements place constraints on the robustness of the data storage process because of the risk of sudden power losses during modification operations. These operations consist of program and erase commands on the Flash device, and their interruption could lead to an inconsistent status in the storage system. Since important system configuration parameters could be stored as normal data, there is the risk of system corruption (critical failure at the next powerup).

Mobile phones and automotive systems have also strict real-time needs: in this case it is not acceptable to wait for an erase/program operation end (usually a "slow" operation) before reading critical information for the system functionality.

This is the reason why Flash architectures (mainly of the NOR type) have been improved with the introduction of independent banks (*dual-bank* and *multiple-bank* devices) that allow them to read from one bank while programming or erasing on another one. Therefore, even if the Flash generation is not helping with advanced features, it is necessary to handle the read-while-write functionality; this means also that on *single-bank devices* the Flash management software has to take care of real-time constraints. This is usually done with an appropriate task *scheduling* policy of the Flash operation, using the resources available inside a *multi-tasking* and *real-time* operating system (usually used in mobile phone platforms).

In the next sections we will present an overview of wireless and automotive applications, focusing on the impact they have on Flash memory management software architecture.

Consumer devices (set-top boxes, printers, PCs, etc.) will not be treated in detail since the role of Flash management SW is less critical. In fact, in this segment the most complicated aspect is the throughput of mass storage units, while the other typical nonvolatile data are simply configuration parameters (e.g., the programs of a set-top box or the BIOS parameters of a PC) without critical requirements in terms of system responsiveness or reliability.

6.2.1 Wireless Market

The mobile phone market has been rapidly changing from the original portable phone applications to the recent platforms offering advanced multimedia services.

In just a few years, from the first devices implementing only voice services or simple textual messaging (SMS, Short Message Service), the market has moved to ever more complex multimedia terminals that combine the original telephonic functions with a set of functionalities typical of a pocket PC.

Figure 6.1 depicts the generational evolution of mobile phones and their functionalities. The acronym 2G commonly indicates the second generation of mobile

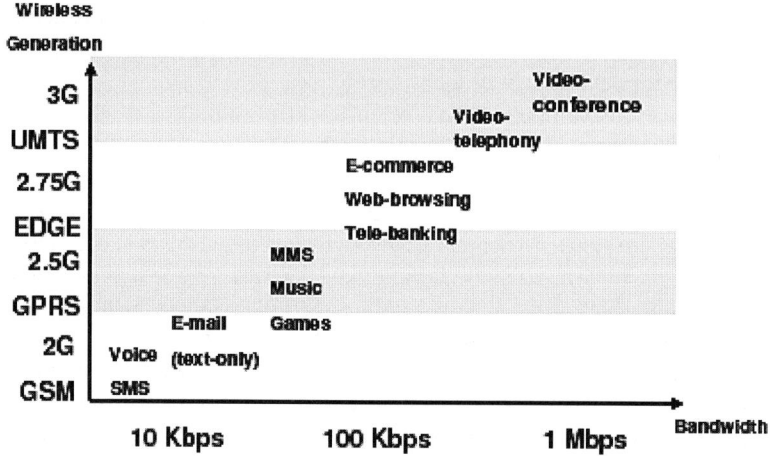

Fig. 6.1 Mobile phone generations, services, and required bandwidth: from GSM vocal services and textual messaging that require only 9.6 Kbps, to the UMTS or WCDMA videophones that need around 1 Mbps

phones for voice communication, and in particular, devices implementing the GSM (Global System for Mobiles) communication protocol that allows textual message exchange (SMS) and textual access to the web. This is the generation that has replaced the first analog version of mobile phones (TACS, Total Access Communication System) offering only voice communication services. The later digital communication protocols have completely changed mobile phone applications both in terms of signal/noise ratio and in terms of communication security and available services.

Between GSM generation (2G) and the most recent family of terminals implementing UMTS protocols (3G), two other intermediate generations have typically been introduced; these have brought important innovations in terms of bandwidth and the resulting available services.

2.5G technology signifies the GPRS (Global Packet Radio Service) terminal era with a wider offer of accessible services like Internet connection with e-mail clients, web browsing, GPS (Global Positioning System), and MMS (Multimedia Messaging System).

2.75G technology includes the EGPRS (Enhanced-GPRS) technology, also known as the EDGE (Enhanced Data rate for GSM Evolution) and CDMA (Code Division Multiple Access) standards. This family is mainly distributed in North America and represents an additional step ahead in multimedia capabilities like video exchange, home banking, e-commerce, wireless gaming, video camcorders, and digital still cameras, thanks to the wider bandwidth.

Only services like videophone calls and video conferencing require third-generation technologies with a bandwidth equal to or greater than 1 Mbps based on the UMTS (Universal Mobile Telecommunication System) standard.

Figure 6.1 also shows the functionalities available with respect to the available bandwidth.

6.2.2 HW and SW Platforms for Mobile Phones

Together with the evolution of transmission technologies there has been an evolution of the HW and SW platforms for signal processing, and in particular of mobile phone platforms. Figure 6.2 shows an example of a 2.5G hardware platform in which there are a radio frequency section; an analog baseband signal treatment section together with a digital signal processing section including a DSP for multimedia applications like video-codec, MP3, and MPEG4 algorithms; a microcontroller for managing the display; a video camera; a SIM card; and a keyboard user interface.

A dedicated hardware section, the memory management unit, is responsible for handling memory peripherals, in particular a volatile memory and one or more non-volatile memories.

The code requiring high execution speeds, like voice signal coding/decoding algorithms or telecommunication management, is usually executed in RAM portions.

Sometimes, additional RAM portions are reserved to preview and temporarily store data coming from WAP pages, allowing the user faster browsing until the connection ends.

Flash memories are typically used to store initial data (mobile phone ID, device configuration, menu, calibration data, etc.), user configurable data (photos, telephone numbers, etc.), and code (executed in place when possible).

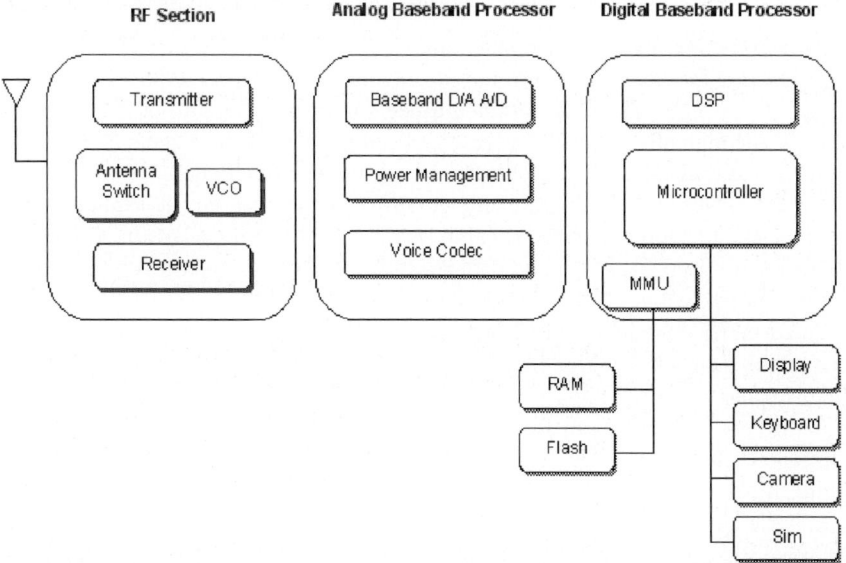

Fig. 6.2 Block diagram of 2.5G mobile phone

The entire storage space is usually partitioned into areas or quotas reserved for each application. For example, the manufacturer fixes the maximum number of telephone numbers that can be saved, the number of photos, the amount of storage for voice dialling, the number of SMS and MMS, and the maximum duration of a movie.

Calibration data need special attention and treatment. Each mobile phone must respect strict and accurate standardization criteria that require a fine tuning of hardware marginalities. These data, together with the phone ID (IMEI, International Mobile Equipment Identification) and other OEM configurations, are stored during production in the factory and never changed.

The size of this OTP (one-time-programmable) area depends on manufacturer needs (e.g., telephone service provider data) and the architecture of the device software (e.g., implementation of interpolation algorithms or storage of the complete set of needed data). In any case, the amount of this kind of data is very small if compared with multimedia data. NOR Flash memories are ideal to store and execute code and OEM data; on the other hand, multimedia information need much more space on Flash and require NAND-based solutions.

It is now clear that new mobile phone architectures have various needs in terms of storage capabilities: not only highly reliable small data areas, such as code, but also large multimedia files. The new HW and SW architectures will be designed in order to merge and treat these differences.

6.2.3 The Automotive Market

In the automotive market, the application range of Flash memories is very wide. They are used for electronic control of the engine, as well as in GPS applications or in security systems. The only common aspects in these different possible applications are the high quality level, fast responsiveness to real time triggers (e.g., airbag control), and critical condition robustness (e.g., temperature range).

For the *power train* there are many and different applications, depending on the different manufacturers and the specific models. In this field, Flash memories are mainly used for code storage and data structures (only a few Kbytes) organized in N-dimensional look-up tables (N between 3 and 4).

These tables contain a set of parameters (calibration data and engine mapping) used by the main software for electronic injection control and all the other functionalities related to the engine.

In the case of applications for passenger security (*central body and gateway* and *instrument cluster, car radio and multimedia, GPS*), Flash memories are used in a similar way but, obviously, with a different kind of data to manage.

In these applications, code and data are very small and do not require high density memories. The typical size of Flash memory is 1 MByte; usually automobile vendors prefer to have a unique solution, including a 16-bit microcontroller, RAM (up to 16 Kbytes), and sometimes a DSP.

New applications in the automotive field include the so-called *l'Infotainment*, today distributed only in top category cars, that brings much more complex multimedia applications and so more complex requirements in terms of data storage.

These requests need to handle big multimedia files, and so size needs and performance needs are increasing. If up to now Flash memories of the NOR type were the only ones present in a car, in the future these different applications will bring more and more NAND Flashes in. The historical NOR Flashes will continue to be used as in the past in integrated solutions, while the new entry NAND Flashes will come into the game in separate systems, usually removable, like memory cards.

6.3 Software Methodologies for Flash Memory Management

The way in which the software manages the Flash memories depends on the type of information which must be stored and on the type of system to which the application is dedicated.

Typical examples of information stored on the Flash are parameters whose size is a few bytes; data with variable size and with different writing/reading speed requirements; files with an associated set of attributes such as name, size, access rights, etc.; and compiled code executed directly by the processor. Each of the described types implies different constraints for Flash management, as we will see in the next paragraphs.

6.3.1 Logical Data Management on Flash

As above mentioned earlier, among the various peculiarities of Flash memories, the most important factor which influences their management from a software point of view is usually the time requested for the sector erase.

Flash memory allows word granularity writings (or page granularity if the device supports a buffer program) for a NOR memory and page granularity (512 bytes–2 kB) for a NAND Flash memory; the erase operation is allowed for both Flash devices only at sector level (typically with a size of hundreds of kBytes). According to the technology used for memory design (if NOR or NAND type), the operation may also require a very long time (from milliseconds for a sector erase on the NAND Flash to one second or more on NOR Flash). Apart from the effect that the erase time has on system performance when the memory is used, a first remarkable impact concerns the management of the device. In fact, it would be possible to use a Flash memory as a normal RAM or EEPROM, on the condition of waiting the erase time of the entire sector in the case of the replacement of a part or the entire contents of it. This, besides not being practicable due to inefficiency in terms of performance, is also inconceivable with reference to the maximum amount of very high but finite programming/erase cycles.

The best solution to avoid this usage limit is to implement a mechanism to manage the data fragments based on the validity of the information. In practice, information is stored in the Flash in segments with the appropriate size related to the type of application and the type of memory. Every segment is associated with a header, i.e., a control information field able to individuate the associated segment in the Flash device and to manage its validity. Validity management means that, at the moment when the software receives a request to update the fragment contents or to delete it, the software is not forced to physically remove data from the device with an erase operation, but simply to consider the related fragment as invalid, executing a logical erase. This kind of mechanism consists of marking a part of the area in the device as invalid or "dirty" and to consider its contents obsolete. We have defined this memory area as "dirty" because it is not possible to use it again for subsequent writes, until a block erase operation is executed. Obviously, the implementation of such a management mechanism also requires the implementation of a mechanism for recovering the dirty space, also known as garbage collection or defragging. The garbage collection algorithms must provide a mechanism to individuate the block which is more opportune to "clean," related to the amount of contained invalid space and to its usage (amount of erase operations and/or time passed since the last erase), in order to prolong the life and reliability of the device. What is easily understandable is that the occurrence of the erase operations is considerably reduced, and therefore the system performance is considerably improved.

Whatever the type of Flash memory (NOR or NAND) and the type of target application, the management policy of the Flash memory remains the same: for both it is necessary to introduce appropriate mechanisms of logic identification, of data logic validity in Flash. In order to implement such a solution one possibility is to use "bit manipulation" feature of flash memories, i.e., the ability to write again in a location previously programmed without generating an error. This last concept must not be interpreted as the ability to program an already written location to any value, but rather the ability to set to 0 the bits that during previous operation were set to 1. Operating this way, it is possible to update "in place" a specific memory location, and in particular the header location, which contains the validity status of the associated fragment.

All the Flash devices usually allow bit manipulation in all the address space (e.g., NOR memories without error correction management) or in part of it (e.g., the "spare" area of NAND memory). An evident consequence of the logic management of the memory and of the introduction of the headers is the usage of part of the address area for control information necessary to allow the system to work. This area used for data management, together with the alignment space of the actual data to the minimum amount of addressable memory, is called *overhead*.

The header of a data very often does not contain validity information only, but also information related to the identification of the fragment, the address of the data area, the size of the data area, and so on. Moreover, the header with its validity field is often used to manage the power loss and subsequent recovery of the consistent status of the data in Flash.

6.3.2 Flash Memory Management and Performance

The main characteristics for evaluating a software for Flash management are the space overhead required for managing a user's data and the system I/O throughput. To optimize the overhead means at least to reduce the amount of space on Flash required to store the control fields, and any other information added to the data payload.

This requirement is particularly relevant in so-called low-end systems, where in order to keep the cost per piece low, small size Flash memories are used. In fact, the software is designed in order to make flexible the minimum quantity of data stored in the Flash, providing an efficient storing method for managing small amounts of data. Often the memories are used for storing both data and executable code. In the latter, at least for NOR memories, user applications are able to be executed directly from Flash, performing what is usually called XIP.

In common language we often hear that some management techniques optimize the Flash space overhead, making flexible the minimum size of the address space. That means that the Flash software tools are able to manage data fragments having variable sizes, and, in some cases, they implement a "smart" usage of the memory space for data smaller than the minimum storage quantity. In many cases, this type of management conflicts with throughput expectations, because it requires a greater computational effort for managing variable chains. If we also add the effort required to implement all the functionalities to keep data integrity in case of power loss, we understand also that basic operations require a lot of operations in Flash (to program the power loss field in the data headers), jeopardizing the throughput.

To improve throughput performance, the data management gives up space overhead optimization either by introducing fixed size fragments or by reducing robustness in the mechanisms that manage the power loss recovery. A strong and consistent management of power loss recovery requires a high number of changes in the control fields, hence a high number of additional programming operations.

NAND memories have architectures and performances, in terms of programming and erasing time, which are especially useful in adopting this kind of approach involving fixed fragments (often called pages) and high throughput. In fact, the real bottleneck for a "fast" Flash software management is the time required to recover the dirty space, that is, the time to perform garbage collection.

There are different techniques for minimizing the impact of this phenomenon during garbage collection. The garbage collection performed in the background (when the system is idle) is the most frequent solution to improve performance. Other techniques include the usage of more spare blocks (up to 10) in case of writing requests when there is lack of free space. In addition to these mechanisms, data management can also implement RAM caching techniques, obtaining a significant improvement in performance. This type of solution, however, requires a much more system RAM, and that means a higher cost in terms of resources.

6.4 Code Storage

The latest generation of mobile phones allows executing applications that the user can "load" on the wireless device. They can be stored on Flash in the form of files and then moved (downloaded) to RAM memory locations in consecutive regions in order to be executed. In this way the processor or virtual machine (in the case of code interpreted as Java) can fetch the next instruction stored in the next physical address. This download may take significant time, especially if the file is large; moreover, RAM space in adjacent locations, may not be enough to store the application. For these reasons, most current platforms for wireless devices often use a tool (a code manager) for storing code or any other object that requires contiguous space and/or direct access through its address. Obviously, these tools are able to manage reading and writing from different applications, always guaranteeing the consistency of the executed code. As mentioned, the NOR Flash is divided into categories: single-bank, and dual- or multi-bank.

In single-bank Flash, the code or portions of it which perform programs or erase commands, cannot reside in Flash, otherwise the processor could not execute the code from Flash after issuing the command. Possible strategies to be followed for this type of Flash are:

- execute the code from RAM;
- store in RAM only the portion of the code that executes the Flash commands. Thus during the period in which the Flash is busy writing or erasing, the processor can still interpret the next instructions stored in RAM.

If the wireless platform is composed of a multitasking/multithreading environment with an operating system using a preemptive scheduler, the second type of solution presents an additional complication. An operating system preemptively manages scheduling by interrupting the execution of the thread/task that is running if another thread/task having higher priority is ready.

So if the device is busy due to an erasing or programming operation, and a higher priority task/thread takes the CPU, it will be unable to read code from Flash.

In order to fix this problem, the data management tool, acting on the interrupt system, disables all the interrupts, assuring that the running task is not preempted.

Moreover, the pending interrupts (usually stored in some pending registers of the hardware platform) can be checked by the low level drivers during the waiting cycle of the performed Flash operation. In case of interruption, the driver routine can suspend the Flash operation (taking care to put the Flash in read mode) and then enable the interrupt system, in order to leave the CPU control to the waiting task/thread without blocking it until the end of the Flash operation. When the scheduler gives again the control to the driver, it will disable the interrupt system, and then it will resume the pending Flash operation. In the dual-/multi-bank Flash devices, the above mentioned problem doesn't happen because it is possible to perform a Flash operation from one bank and continue to execute code from another one.

6.5 Common Problems in Flash Memory Software Management

6.5.1 Wear Leveling

Flash memories have a limited life time. In particular, with alternating write and erase cycles, the Flash blocks tend to become unreliable, showing cells which lose the capability of changing their state. This phenomenon appears after a number of cycles, which usually ranges between 10,000 and 100,000, according to the device technology and characteristics. Consequences of this phenomenon include the impossibility of using the entire block to which the spoiled cells belong (besides, further cycles could spoil more cells); and, in the long run, the necessity of replacing the spoiled device.

It's obvious that the importance of countering this undesirable behaviour of the devices depends on the type of application: for example, in the Flash memories used to store a PC's BIOS, the number of writing/erase cycles of the blocks is limited; on the other hand, in more dynamic applications like cameras or multimedia players, it is necessary to extend the life of the device as much as possible. In any case, while the spoiling of blocks is irremediable, it is possible to slow down the phenomenon via software with "wear leveling" techniques. These techniques are based on two considerations.

First, in reality, not all the information stored in the device change with the same frequency: some of them are updated often, while some other remain unchanged for long time—at the most, for the whole life of the device. It's obvious that the blocks containing frequently updated information are stressed with a large number of write/erase cycles; while the blocks containing information updated very rarely are much less stressed. Besides, it is possible, with acceptable costs, to have Flashes with a surplus of space (typically one block named "spare block") compared to the space strictly necessary to store data. The wear leveling technique exploits the extra space available, moving the information around among the physical blocks of the device, in order to average the number of erases of each block, extending the life of the device.

For example, let's suppose we have a device with 10 blocks, one with information updated 100 times in a day, and the others updated once a week. If the device has a life limited to 100,000 erase cycles, after about 1,000 days (which means less than 3 years), the most heavily used block could start to be stressed, while the other blocks could last about 2,000 years! With wear leveling on 11 blocks, the erases would be distributed among all the blocks regularly, with an average of less than 9 erases for each block, extending the life of the device more than ten times. Unfortunately, in reality, it is not possible to apply the wear leveling in a very accurate way, and so there will always exist some blocks erased more often than others. In order to allow the information to move around among the blocks, it is necessary that they are not referenced using their absolute position on the Flash; but that a procedure be used (in hardware, software, or firmware, according to the cases) which can associate the sought information dynamically with its physical position on the device. The description of the various techniques to obtain this result (as well as the pros and cons of each of them) is beyond the scope of this chapter. What distinguishes the different wear leveling algorithms is, essentially, the logic used to choose the block to erase. This can be based either on the total number of erase cycles of each block during its life (which in this case must be stored on the device itself), or on a rotational logic, possibly considering also the amount of space which can be recovered by erasing the block.

6.5.2 Power Loss Recovery

In all data storage applications it is necessary to guarantee the correctness of the data as much as possible. The main cause of data loss is the loss of power for the device during writing or erasing operations. In fact, because of their nature, the memories are not able to identify interrupted operations and to complete them. This means, for example, that very big amounts of data can be stored only partially; or, on the other hand, that data no longer useful—because replaced by newer data—still resides on the device and is not distinguishable from the original data. In addition, code replace operations interrupted by power loss may lead to very bad effects, until the entire system is no longer working. For these reasons it is necessary to introduce a data storage policy which is able to identify spoiled information and, where possible, to recover it.

It is important to note that the concept "recovering information in case of power loss" is not unequivocally defined. Depending on the application, the data management policy, and the amount of available resources to manage power loss, the type of recovery can vary. One can move from a policy aimed at guaranteeing that each single operation is either interrupted or completed at the startup of the system, to a policy able only to guarantee that no data is corrupted except what is involved in the current operation, to a policy which simply guarantees the coherence of the device structure but nothing about its contents. Similarly, there are differences in the ways which power loss recovery is implemented. The main techniques can be classified as follows:

- techniques based on status bits;
- transactional techniques;
- journaling techniques.

6.5.2.1 Techniques Based on Status Bits

In these solutions, all data is associated to a set of bits, whose size depends on the application's necessity. These bits are used to monitor the time evolution for the validity of the information. A very simple example would be the following set of data:

111 data not used (initial value)
011 writing data
001 valid data
000 invalid data

During the ordinary system execution, every information follows the sequence $111 \rightarrow 011 \rightarrow 001 \rightarrow 000$. At the startup, the system controls the status bits of all the data and, if it identifies interrupted operations, it either completes or discards them. In the previous case, each item whose status bit set at 011 could be either incomplete or incorrect, and then it must be declared as invalid, converting its status to 000. It is important to point out also that the progression of the status bits must be carefully monitored and defined a priori; for the previous example, in the case of power loss, the status bits change according to the following sequence: $111 \rightarrow 011 \rightarrow$ (power loss) $\rightarrow 000$.

Let's suppose now that a new power loss again interrupts the transition $011 \rightarrow 000$; the status of the data could be 011 or 000 or even 001. At the next startup the system would wrongly identify the valid data, making the entire procedure useless. A correct sequence in this case would have been the following:

$$111 \rightarrow 011 \rightarrow \text{(power loss)} \rightarrow 010 \rightarrow 000$$

In fact, the passage through the intermediate state 010 allows the system to identify the data unequivocally as "invalidating." The complete state set becomes the following:

111 data not used (initial value)
011 writing data
001 valid data
010 invalidating data
000 invalid data

Generally, the set of values and the state transactions must be defined in a way that each state change implies the transaction of one and only one bit.

6.5.2.2 Transactional Techniques

These kinds of techniques are usually used when each user operation translates to a set of simple operations on the Flash—as in the case of file system operations.

In this case, the operations to be executed on the device, instead of being directly executed, are stored on the device together with all the additional information necessary to complete them. Only in correspondence with a specific instruction from the higher SW layer will the Flash management software start physically executing the operations; in this way, in case of power loss at reboot, the system completes the possible set of operations prematurely interrupted.

The same philosophy is used also by systems which operate in the opposite way: executing the operations immediately and, if necessary, cancelling them in case of power loss before confirmation. The different techniques belonging to this category distinguish between the sets of simple operations recognized, the ways the information about the operations is stored, and so on.

The advantage of this technique compared to the previous one is that the required space to manage the transactions is necessary only during the operations, while the techniques based on the status bit require an additional space permanently for each item of data; but these techniques result in a simpler implementation and are often more efficient, requiring a smaller number of operations to program the Flash.

6.5.2.3 Journaling Techniques

These techniques, applied mainly in the file systems field, tend to exploit the advantages of the transactional approach, limiting as much as possible the performance drawbacks, but sacrificing the guarantee of consistency of the single items of data. In these techniques, the transactions are not applied to all information contained on the device, but only to so-called metadata, that is, all the additional information used to identify or to recover specific information. Typical examples are the file name, its size, and the directory where it resides.

The big advantage for these techniques is that, by limiting the transition only to some information stored on the device, they succeed in giving more robustness to the whole system; in fact, in case of a power loss, at most only the files on which writing operations are ongoing would be corrupted, files which can be easily identified. Besides, the progression of these techniques has made the update phase of the original metadata not strictly necessary, as a result of completing the whole sequence of operations. Briefly, the current techniques manage each item of metadata as an original version available in Flash, followed by the next updates (journaling copies) still stored on Flash, reachable through a pointed list. Going through the list, it is possible to read the updated metadata. Further improvements have then allowed making the scanning of the list more efficient.

6.5.3 Garbage Collection

Garbage collection is the procedure which allows the efficient use of space on the Flash. This procedure has two different goals:

1. to recover the space occupied by data which is invalid, because it is obsolete or has been spoiled due to a power loss;
2. to compact the space occupied by the still valid data, in order to make the management of the free space more efficient.

There are two different policies for garbage collection: the first is used by the applications able to localize the particular information independently by its physical position (e.g., because data are localized with a name, as for the files on the disk), and then able to move them dynamically inside the device; on the other hand, the second is used with those applications (e.g., devoted to managing code executed directly from the Flash) in which the data may be extended on contiguous blocks, or the positioning of which, either absolute (the address) or relative (the offset between information fragments) must be preserved.

In both cases, garbage collection is based on the use of a spare block.

6.5.3.1 Dynamic Data

For this type of solution, which is more flexible and efficient, the block used as spare changes during execution. The general procedure followed for garbage collection is the following:

- a block to reclaim is chosen;
- only the valid data are copied to the spare block;
- the spare block is declared valid and at the same time the reclaimed block is declared invalid;
- the reclaimed block is erased and will be used as the spare block in the next cycle.

Whenever the space recovered by this method is not enough, the procedure will be repeated on a new block, until all the available space is recovered. With this solution, every cycle requires a single block erase; the choice of the block to reclaim is based on considerations about:

- the amount of space recoverable with garbage collection;
- the number of erase cycles for the block (wear leveling).

The influence of these specific parameters on the choice of the block to reclaim changes each time, based on considerations linked to the specific application, varying from solutions which neglect wear leveling, to solutions which, for particular cases, can proceed with garbage collection of blocks without any dirty space.

6.5.3.2 Fixed Data or Data Distributed over More than One Block

In this case garbage collection proceeds in a different way. It uses a specific fixed block of the device as the spare block, without changing it, thus making inapplicable the wear leveling procedures. (the spare block will always be the most erased and then it will be the first to deteriorate).

Briefly, the procedure is the following:

- the first block to reclaim is chosen;
- only the valid data are copied from the reclaim block to the spare block;
- the cleaned block is erased and its contents will be restored from the spare block.

The spare block itself is erased to be ready for the next cycle. As mentioned above, in these solutions the spare block is erased more times than the other ones. The case in which data can be extended on more than one contiguous block is harder to manage. In this case the procedure is like that previously described, but it is repeated to preserve the contiguity of the data on all the blocks of the device, with extended time required.

6.5.4 Bad Block Management

In the NAND devices, it is possible that single Flash bits are blocked to a fixed value and cannot be changed. This can happen either during production in the factory, or in a later stage, during Flash usage. Obviously the usage of spoiled blocks is not advisable because of their unreliability. In order to mark invalid the spoiled blocks, a specific position inside the spare area of the block is used. The sixth byte (or the first word, with 16-bit memory) of the spare area of each block is assigned to individualize a spoiled block according to a de facto standard (if different from 0xFF or 0xFFFF). The possibility is neglected that a block may be spoiled in a permanent way by fixing to 1 all the bits of the word or byte assigned to keep the reliability status of the block. When one or more invalid blocks are individualized, there is the necessity to manage them in an adequate manner. There are two solutions to avoid the usage of invalid blocks.

6.5.4.1 Skip Block

This solution affects the flash blocks allocation itself: instead of marking a block as invalid, the algorithm simply allocates only the valid blocks. In practice, in this solution, the 5th block will not be the 5th Flash block, but the 5th valid Flash block. Whenever a new block is spoiled during the Flash's life, this will determine the need to remap the spoiled block and all the next ones.

6.5.4.2 Replacement Block

In this technique the user cannot read/write to all the Flash blocks, but only to a fraction of them (e.g., 80%). When one of the visible blocks becomes invalid, the software starts using one of the others in place of the first, in a completely transparent way. Obviously, all the information needed to individualize the real physical block associated with the data will be stored inside the Flash (typically in the spare areas), starting from a logical block number. The specific solution that is adopted will change depending on the case.

6.5.5 Multilevel Flash

In multilevel Flash memories, the ability to store more than 2 logic states for a cell has allowed doubling the actual capacity of the device. Unfortunately, this positive effect has negative consequences, especially in the case of NOR memories. In this type of memory, in fact, it is generally possible (except in some specific cases: see ECC) to program some bits separately from the next ones, allowing so-called bit manipulation. This behaviour is very useful in the case of devices with a power loss problem, because it allows the use of single bits to store the progression of the operations held on the device (as in the case of power loss recovery). Unfortunately, this behaviour is not possible in the case of MLC devices. In these devices, in fact, the voltage stored in a cell does not identify the logic state of one bit, but two bits (Fig. 6.3, left side).

The consequence of this logic is that in the case of power loss, the manipulation of a single bit may interfere, changing the status of a different bit. For example, during the passage from the status "11" to "01," the cell could transit to the "10" state unintentionally. In this case not only is the system not generally able to identify the power loss event, but even if it is able to identify it, it is not able to restore the correct state of the cell, because it needs to raise the cell voltage without erasing the block which it belongs to. The only possible solution is to give up the advantages of multibit cells for all the information which needs bit manipulation. In practice, only the extreme states of the cell are considered valid (Fig. 6.3, right side), while the remaining states are considered only temporary ones, crossed during the transition $11 \rightarrow 00$. At the system startup, cells may exist in the intermediate states, but only if a power loss has occurred during the transition to the "00" status can that transition be completed. It is obvious that, at least in the case of MLC devices, reducing the number of cells needed for bit manipulation as much as possible allows the recovery of useful space.

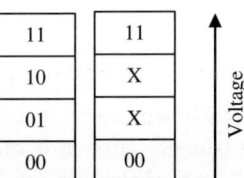

Fig. 6.3 Logic state and threshold voltage in multilevel Flash cells

6.5.6 *Error Correction Code (ECC)*

NAND type Flash memory (and some NOR multilevel devices) are not 100% reliable during writing operations. In practice, capacitive effects due to the presence of nearby cells, and the need to manage very small voltage differences on a single transistor, may mean that some cells do not reach the desired status, or that they do not save it for a long time. The phenomenon may typically involve 1 bit per 10 billion programmed bits, that is to say, for 1 Gb flash device, 1 error may occur after 10 times full device program operations.

In order to guarantee data correctness during the reading phase, error correction codes (ECC) can be used. In practice, besides actual data bits, some additional bits are stored on the device; these bits are properly calculated starting from the original data, and they allow one to identify possible errors and to correct them. The choice of technique used depends on the size of the minimum programmable unit (from a few bytes up to several hundreds). For example, in the case of NAND memory, typically one can use codes which add 22 bits for each 2048 data bits, allowing the algorithm to detect 2 error bits and to correct errors on a single bit. Big differences exist in the way ECC management is implemented in NOR and in NAND Flash memories. In particular, in the NOR memories, the ECC calculation and its application to correct errors is fully implemented inside the device, implying some constraints (quite strong) on the ability to program the device. The necessity to store the ECC code implies that the device cannot be programmed at the bit level like the "classic" NOR, but must use as a minimum unit the fragment size on which the ECC is calculated. If, therefore, programming the same region of the device more than once was allowed, it would be necessary to calculate the correction bits again, and to store the new value, which is absolutely independent of the previous one. The new storage is not possible because of the well known physical constraints on the $0 \rightarrow 1$ transition of Flash memories.

In NAND memories, on the other hand, the ECC value in each spare area of each block is stored next to the actual data values. The actual calculation of the correction bit during the storage phase and its verification during the reading phase are thus referred to external software or, when performance is fundamental, to a dedicated circuit external to the device.

6.6 SW Architectures

6.6.1 *Data and File Management*

All the items described in the previous sections give an idea about the complexity of the constraints which must be managed by tools for Flash memories. All these difficulties can be managed only through an efficient software design in terms of modularity and encapsulation. Typically these software products are organized by software levels and each level implements a specific part of the software, often

Fig. 6.4 Typical Flash
management software stack

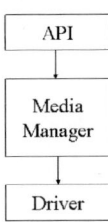

contained in a module of the system. The module provides software interfaces to
higher levels, which hide all the details of the implementation. This kind of hierarchy brings advantages in terms of manageability, reusability, and complexity.
From the user's perspective, these tools typically make their functionality available
through subsets of functions, APIs (Application Programming Interfaces), which
are used by external applications. So in general, there is an interface module which
implements the APIs, as shown in Fig. 6.4.

Close to the hardware device, there is the software driver, which is able to manage
the Flash devices in order to handle operations like program, erase, and read in a
very simple way. In addition to the basic operations (standard command set), the
driver is also able to perform advanced operations (extended command set) like fast
programming, protections, and so on.

Often, inside the driver, there are routines for automatic Flash recognition through
querying of the JEDEC-compliant CFI (common flash interface), which stores information like geometry, manufacturer, model, the commands set, etc. Using this
technique, it is possible to change Flash devices without modifying the software,
thereby improving the software portability.

Between the API and the driver levels there is the heart of the software system, the media manager. It implements strategies for storing the data as well as all
the features which typically must be managed by this kind of tool (wear levelling,
garbage collection, etc.). In order to understand the media manager characteristics,
it's important to understand that when we talk about data, in general, we are referring
to objects with a unique identifier, with a fixed size, and within which you can carry
out operations like writing, reading, and deleting. Typical data are: configuration parameters of a mobile phone (volume, display brightness, etc.), a short text message
(SMS), a phone book entry, etc.

A file, instead, is a special type of data which owns a unique identifier represented by its name and has a number of attributes that determines its access rights.
Photos, programs, and audios are typically stored as files. In general we talk about
a file manager or a data manager, depending on the type of objects which the media
manager operates on.

In accordance with this nomenclature, the top layer of an API will provide either
file management (read, write, delete, open, close, edit access rights, create folders,
etc.) or data management (read, write, delete) functionalities. In many cases it is
possible that the API layer provides both data and file management functionalities.

In principle, a file can be considered as a pair of two sets of data, i.e., the file content and the file attributes (name, rights, folder, etc.), so it is possible to provide file management functionalities even if the media manager consists only of a data manager; for this reason, in this chapter we will simply speak about *data*.

6.6.2 The Data Manager

In the previous section it has been mentioned that one of the objectives of a Flash management tool is its efficiency in terms of storage speed and data management policy. The Flash memory is the bottleneck of the whole system because, although its reading performance is comparable to that of RAM, it requires much time for programming and especially erasing blocks. It is unthinkable that during the writing of data, if you needed to free invalid space, the user application would have to wait for the physical erasure termination of one or more blocks.

The solution which usually is implemented to face up to this problem requires the usage of multiple cooperating processes (multitasking), as shown in Fig. 6.5. The system of data management in Flash memory is composed of at least 2 tasks with different priorities. The highest priority task—the foreground—provides the API, while the lowest priority task—the background—is the data manager core. These two processes communicate with each other through a data structure—the queue—which stores the requests for creation and/or modification of data. When a user needs to operate on specific data, he will call the specific API, which will insert an appropriate command into the queue, without waiting for its termination.

While the system is idle, the background process starts its job by taking one command from the queue and executing it on the Flash, also managing the invalid space and recovering it when needed.

Adopting this architecture, if during a Flash operation the foreground process is awaked by an API call, the user application takes control of the CPU by blocking the data manager routine execution with a maximum latency not greater than the time required for a context switch. Differently from the write operations, the read operations must be executed within the constraints of promptness, but at the same time must be consistent with the queue and the Flash contents, so they are

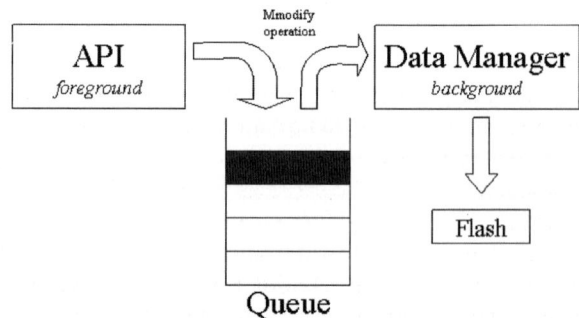

Fig. 6.5 Multitasking

performed directly without passing through the background and reconstructing the actual content of the data.

For example, if the first 2 bytes of a given set of data stored in Flash are "AB," and the queue contains a write operation of the second byte of the data from "B" to "C," a reading operation of the first two bytes is supposed to return "AC."

The Flash memory in this case becomes a shared resource between two tasks (foreground and background), and since it is a static machine which during program or erase operations does not provide data content related to a required location, the software must provide a mechanism to synchronize the tasks. This kind of mechanism, by suspending and resuming the Flash operations, pauses the foreground task until the device is not ready to be read.

Although the multitasking solution improves the overall performance of the system, it brings more problems in case of sudden power interruption (power loss) typical of battery-powered devices such as phones. In fact, since the requests for data modifications are stored in the queue (volatile memory), an interruption of power inevitably implies the loss of updated data. To reduce the risks caused by the power loss, the queue must contain the right number of operations by balancing the performance and the robustness. This can be achieved by periodically flushing the queue or by creating extremely efficient background routines.

To improve the efficiency of the system, sometimes the garbage collector (a routine to free invalid space) is executed in a separate task having a lower priority than the background.

Using this option, when the system is idle and the queue is empty, the garbage cleaner takes control of the the CPU and starts the cleaning of the Flash, either until the space does not reach a threshold (generally configured by the user), or until a higher priority task (e.g., an API) has been awoken. If the system is configured at its best, the write operations are faster because they will not be forced to call the cleaning routines.

Both multitask solutions (with or without garbage collection as a separate task), unlike single task solutions where there is neither a command queue nor a synchronization module, suffer from a serious problem linked to the results of operations required by the API. Since the success of an operation is known only when the command is really executed on Flash, the foreground process usually assumes that the operation has been successfully executed, giving a sense of promptness as a response to the application which is using the data manager.

The result of the operation will be verified through a periodic check of a proper common error channel that can be, in general, a messages queue containing the results of each operation or simply a RAM structure containing information on the failures.

Often data management systems provide some secure write APIs in which the foreground task is locked until the command has been executed on the Flash. In this case the API returns the true result of the executed operation.

Figure 6.6 shows the described architecture in which the synchronization module synchronizes all the activities performed by the system tasks, by using the features provided by the operating system (semaphores, mutex, events, etc.).

Fig. 6.6 Data manager architecture

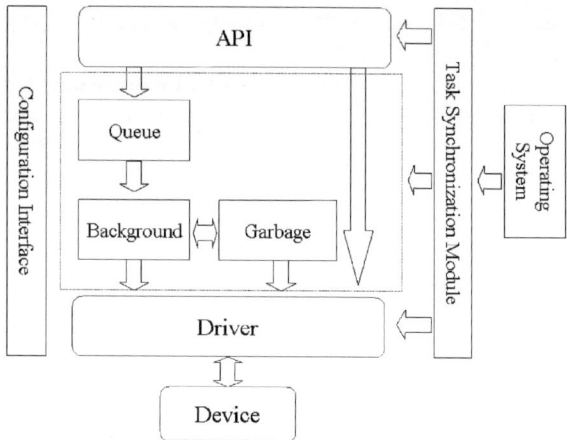

Such architectures are typically adopted in low-end wireless systems, which are capable of managing small size sets of data, and which incorporate proprietary OS solutions characterized by tiny kernels and with no file system capabilities. Vice versa, high performance wireless systems (*hi-entry*), which generally adopt operating systems with a native file system, often use tools to manage Flash memories that export typical disk-like functionalities (disk emulators) to the upper layers in a fashion transparent to the user.

The next section describes both the philosophy behind this kind of tool and a proposal for one of the possible architectures, going into some of its details.

Often, in order to make porting of the software to different operating systems easier, the interface to the synchronization functionalities is concentrated within a single module (often called OSAL—operating system abstraction layer) that will be modified when necessary.

The block configuration is the module containing all the configuration parameters needed to enable or disable some system functionalities, like the threshold for the garbage collection, the size of the storing area, etc.

Obviously, the architecture described in Fig. 6.6 is general enough that customization in the direction of increasing or reducing its complexity is always possible.

6.7 Disk Emulator

The disk emulators constitute a category of tools for data management designed to hide all the characteristics and constraints of the Flash, letting the Flash appear as a hard disk. This means that above the disk emulator, the higher software levels see the Flash as a consecutive subset of sectors (typically having 512B size), which can be read and written to without managing any problems like garbage collection, wear leveling, and so on. In this way any file system designed for hard disks can be used

Fig. 6.7 Disk emulator

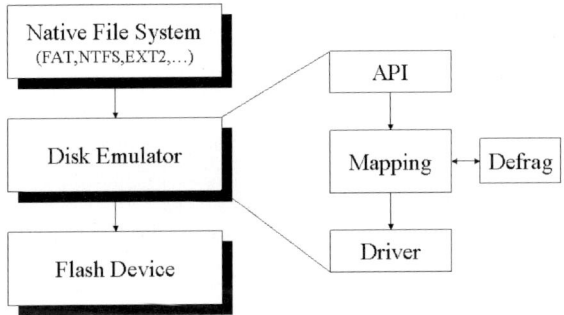

on Flash without any modifications (see Fig. 6.7). Each sector, as on the hard disk, has its own unique identifier (sector number), and the file system refers to it by using that identifier. When the file system needs to write a file onto a hard disk, it will send a series of writing commands to the driver of the disk, specifying the contents and location of the sectors (in terms of the number of the logical block). The driver of the hard disk translates the logical address into a physical address in terms of a CHS (cylinder, head, sector) address and executes the write operation.

In order to emulate the hard disk behaviour, these tools split the Flash memory into physical sectors and erase units. One erase unit is the smallest Flash unit that can be cleaned with an erase operation, and it contains a preset number of physical sectors, depending on the size of the logical blocks addressed by the file system.

When the file system addresses a logical block, the disk emulator translates this logical address into a corresponding physical address for the sector on the Flash. The real issue to manage is that, while on a disk a rewrite operation is possible by changing the sector's content, in a Flash memory the contents of a physical sector can be changed only after the first content is erased.

That is why a validity field is associated with each sector, together with a field containing the number of the logical sector which it refers to. When the file system asks to rewrite the content of a logical block, the corresponding physical block will be declared invalid and the new content will be written to an empty sector which will be assigned the same logical identifier. This approach will cause the higher software levels to "think" that the Flash was written to "in place," while actually the sector was written to a different place. Of course, during its usage the disk emulator will create a big amount of dirty space (sectors made invalid), which can be recovered by executing an opportune garbage collection procedure (a defrag).

The logical to physical translation (mapping) generally is done through an internal table (a mapping table, shown in Fig. 6.8), which very often is kept in RAM to speed operational access to the areas. The mapping table is loaded at boot time by scanning the whole Flash. In this procedure, each logical identifier of the valid sectors (assigned the first time by the format procedure) is read; and for each identifier, the mapping table stores the physical address where the sector is located.

Generally, disk emulators are used in applications characterized by complex operating systems (Linux, Windows, WinCE) with native file systems running on

Fig. 6.8 Logical/physical
mapping

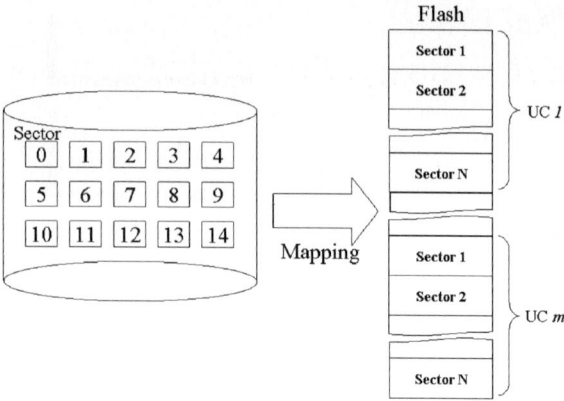

block devices. Typically, these tools are single task tools, because all issues related to buffering and synchronization between applications and high-level Flash are
managed by the operating system. That is why the software architectures are very
simple, as shown in Fig. 6.7. They are essentially characterized by three levels. The
first level, the API layer, provides the functionality for formatting and initializing
the partition (all physical sectors), as well as the functions for writing and reading
logical blocks. The middle layer performs the logical to physical translation and is
responsible for the defragmentation. Finally, as with the other architectures we have
already examined, there is the driver, which offers a set of features to carry out basic
operations like programming or erasing on Flash.

References

1. STMicroelectronics, AN1821 - Garbage collection in Single Level Cell NAND Flash Memories, http://www.st.com/stonline/books/pdf/docs/10121.pdf
2. Red Hat, JFFS, D. Woodhouse, http://sources.redhat.com/jffs2/jffs2.pdf
3. Wikipedia: Journaling file system, http://en.wikipedia.org/wiki/Journaling_filesystem
4. Intel, FDI User Guide, ftp://download.intel.com/design/flcomp/manuals/fdi/chapt11.pdf

Chapter 7
Error Correction Codes

A. Marelli, R. Ravasio, R. Micheloni and M. Lunelli

7.1 Introduction

"HiTech" devices, which permeate our life, carry out an infinite number of operations thanks to the information stored inside. The memory contained in these devices, like the memory of human beings, is neither infallible nor eternal. We all live with these devices, accepting the risk that, every now and then, they might fail. It is up to electronic systems designers to evaluate, for every specific application, an "acceptable" error probability. The acceptable error probability in the control unit of an automobile engine or of the car's ABS is, for obvious reasons, very different from the acceptable error probability for the memory of a video game; and the impact of this reliability on the cost of the device is surely different.

The physical means of storage allows the realization of a wide range of memory devices that vary in dimension, speed, cost, consumption, and probability of error. It is always the designer's responsibility to choose the most suitable kind of memory for the system. Error correction codes (ECC) are a very powerful tool to bridge the gap between the available probability of error guaranteed by the physical means of storage and the probability of error required by the application.

In this chapter we will briefly review the main factors describing the probability of error in Flash memories, and how time modifies them. A simplified explanation will clarify the fundamental theoretical aspects of Hamming codes and will introduce the effects of one or more error corrections on a device's reliability. We will therefore present three different situations with their respective theoretical

A. Marelli
Qimonda Italy Design Center, Vimercate, Italy

R. Ravasio
Qimonda Italy Design Center, Vimercate, Italy

R. Micheloni
Qimonda Italy Design Center, Vimercate, Italy

M. Lunelli
University of Milan Bicocca, Milan, Italy

R. Micheloni et al. (eds.), *Memories in Wireless Systems*,
© Springer-Verlag Berlin Heidelberg 2008

treatments in detail. In the first of these cases a Hamming code is applied to a 2-level NOR memory; in the second case a Hamming code is applied to a 4-level NOR memory; and finally a Hamming code for blocks of great dimensions is applied to a NOR or a NAND memory.

7.2 Error Sources in Flash Memories

The ability of a Flash memory to preserve information is linked to the ability of the floating-gate (Fig. 7.1) to keep a charge, inserted through a write operation, for a long period.

Through erase and write operations, negative charges are inserted into or taken from the floating-gate, in order to get two distributions.

Figure 7.2 shows a variable (current or voltage) indirectly linked to the charge state of the floating-gate. The cells on the right of a reference value (REF) are assigned to the logical value "0," the others to the logical value "1." As time goes by, the floating-gate charge recovers the conditions of equilibrium, losing the stored information. The read and write operations induce a stress on the cells that tends to modify their states. This is intensified by the fact that the cells are organized in a matrix.

In Flash memories the first error source is the cycling, i.e., the sequence of program and erase operations. These write operations require the use of intense electric fields on the oxides of the cells, for example, to inject or to extract charges from the floating-gate. Such operations degrade the capability of the oxide to keep the gate isolated, and therefore also degrade the capability of the cell to preserve the information in time or under the stress induced by the readings and the writings.

Fig. 7.1 Representation of the structure of a floating-gate memory cell

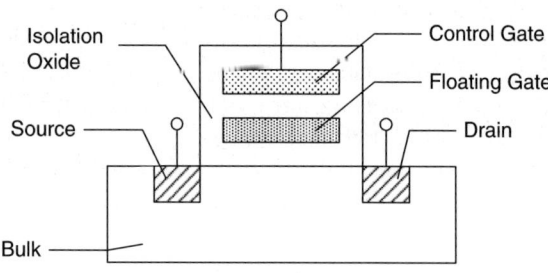

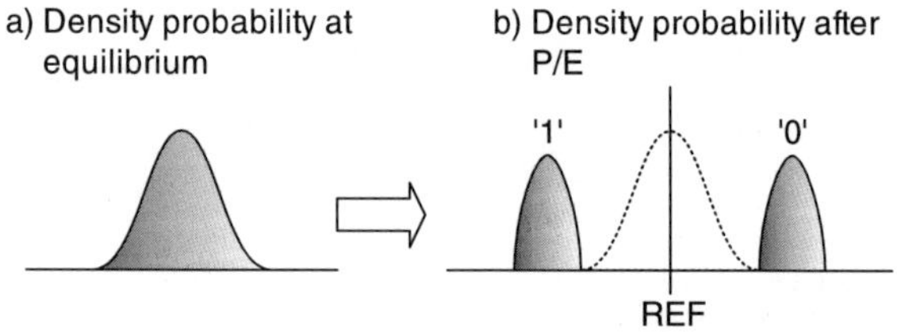

Fig. 7.2 Representation of the distribution of the reference variable: (**a**) at equilibrium and (**b**) after the erase and program operations; REF represents the value of the variable discriminating between the states "0" and "1"

The architecture of the memory is similar to the one described for the NOR SLC memory, in which the parallelism of 128 bits of data is achieved starting from 64 matrix cells through decision circuits (sense amplifier) that extract two information bits from each cell (Fig. 7.18).

The minimum number of parity symbols (cells) to be used is given by the Hamming inequality (Eq. (7.3)), which for $q = 4$ and $t = 1$ is reduced to:

$$4^{n-k} \geq 1 + 3n \tag{7.39}$$

For k equal to 64 symbols, the minimum number of parity symbols is equal to 4. The code used is therefore a Hamming code $C[68,64]$ in $GF(4)$. All the considerations made for the NOR SLC case can also be applied to this code using operators in $GF(4)$. Since the necessary operations to execute Hamming $C[68,64]$ procedures in $GF(4)$ require operators in $GF(2)$ (XOR and AND), it is useful to translate the code in $GF(2)$. This can be obtained simply by converting the data vectors and the matrices from $GF(4)$ to $GF(2)$, according to the decimal representation of the elements of $GF(4)$ in $GF(2)$, as in Table 7.2. For the parity matrix, any row pair of $P_{GF(2)}$ corresponds to each row of the matrix $P_{GF(4)}$.

In $GF(4)$ it is very easy to get the matrix P: We just have to select 64 different rows, using 4 symbols, different from the null row and different from the identity matrix rows. Basically, to optimize the critical timing path and the area of the parity calculation block, it is convenient to search $P_{GF(2)}$ as already described for the bilevel NOR case. Given r_i and r_j, two rows of the parity matrix, the searching method of the matrix $P_{GF(2)}$ must take care of the following constraints:

- Each row r_i in $P_{GF(2)}$ and the sum of each pair of rows $r_i + r_{i-1}$ with odd i must be different from the vectors with one or two 1.
- Each row r_i in $P_{GF(2)}$ and the sum of each pair of rows $r_i + r_{i-1}$ with odd i must be different from every other row r_j in $P_{GF(2)}$ and from the sum of every other pair of rows $r_j + r_{j-1}$ with j odd.
- Each row r_i in $P_{GF(2)}$ and the sum of each pair of rows $r_i + r_{i-1}$ with odd i must be different from 0.
- Minimize the base 2 logarithm of the maximum number of "1" in each column of P, to minimize the number of levels of the adders tree.
- Minimize the number of adders required for the calculation of the parity, by minimizing the number of "1" in the whole matrix P and by maximizing the number of the common terms among columns.

Table 7.2 Decimal conversion from $GF(4)$ towards $GF(2)$

GF(4)	0	1	2	3
GF(2)	00	01	10	11

We leave to the interested reader the exercise of identifying how many cells must be added to take the minimum distance to 4 and to identify the conditions to determine the optimum $P_{GF(2)}$.

7.8 Algorithmic Hamming Code for Big Size Blocks

In this section we examine some solutions adopted in systems composed by a microcontroller and by a NOR or NAND Flash memory, without embedded ECC, and with a reliability considered not sufficient by the designer. In these cases, to increase the reliability of the data of the stored code, the controller applies an encoding to the data to be written in the Flash memory and a decoding on the data read and transferred to the system memory. These operations should use the minimum system resources, both in terms of ROM and RAM memory and in terms of calculation resources (Fig. 7.19).

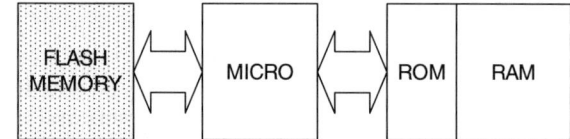

Fig. 7.19 Reference architecture

Suppose we have an application where the memory is logically divided into 512-byte blocks and where, in order to reach the required reliability, we want to correct an erroneous bit and to detect two errors. The Hamming inequality (Eq. (7.3)) requires at least 13 bits plus one for the extension of the code (Eq. (7.4)). The number of parity bits required is extremely reduced, but the parity matrix P consists of 4096×14 bits. Unlike the previous applications, where the parity calculation time had to be performed in a few nanoseconds and where the data to be decoded were immediately all available, in this application data are moved from the Flash memory to the data memory byte by byte (and vice versa), so that parity and syndrome computation cannot be immediate. It is necessary to use a method able to perform the calculation of the parity and of the syndrome without storing the parity matrix in memory.

As already pointed out:

$$s = eH^T = (e_D, e_P)(P^T, I)^T = e_D P + e_P \tag{7.40}$$

That is, the syndrome is a function only of the error. The matrix $H^T = (P^T, I)^T$ contains the information regarding the association rule for each syndrome with the respective error position.

The flow of 512-byte data can be represented as a matrix M with size 8×512 (Fig. 7.20). Therefore, the values corresponding to the vectors of the matrix P (Fig. 7.21a) can simply be the coordinates x and y of the position of each bit in

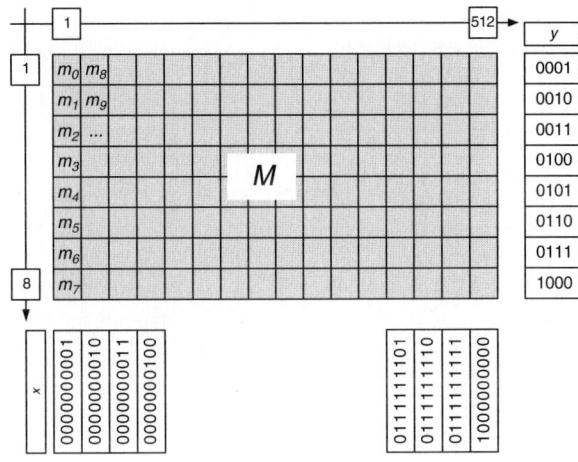

Fig. 7.20 Representation of the matrix M systematically built on the basis of the data flow (m_i) going through the memory; the sorting indexes x (1–512) and y (1–8) univocally identify each bit of the matrix M and are therefore suitable for building the parity matrix H

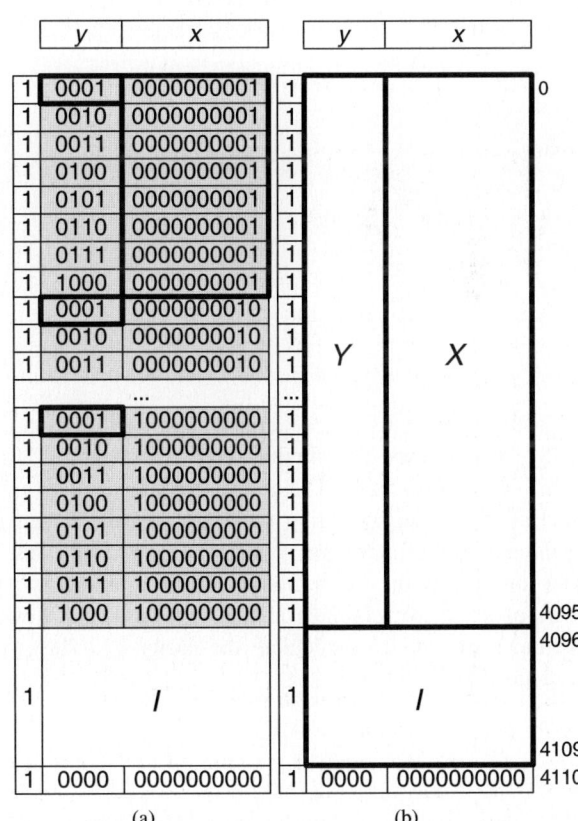

Fig. 7.21 Representation of the matrix H. The sorting indexes x (1–512) and y (1–8), univocally identify each row of the matrix H (**a**). For the calculation of the parity and of the syndrome, the matrix H is seen as the composition $((Y, X)^T, I)$ (**b**).

Fig. 7.22 Representation of the matrix M systematically built on the basis of the data flow going through the memory; the sorting indexes x (0–511) and y (0–7) univocally identify each bit of the matrix M, and are therefore suitable for building the parity matrix H

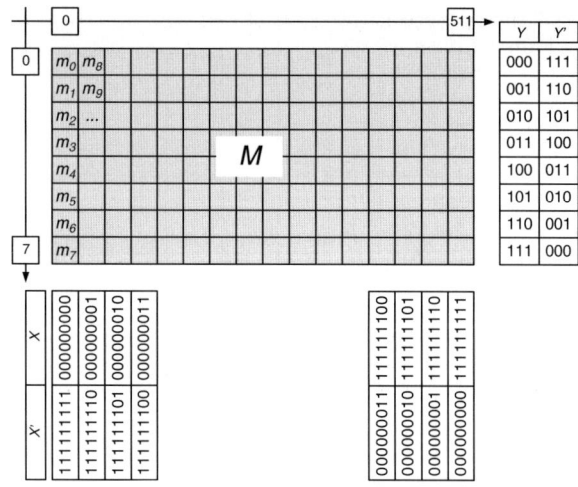

the matrix M. Note the lack of the value "0" for the indexes x and y to avoid the null vector and the vectors of the matrix I in P.

Being X and Y the two matrices composing P (Fig. 7.21b), the parity vector p can be calculated as:

$$p = mP = m(Y, X) = (mY, mX) \tag{7.41}$$

$$mX = \sum_{j=1}^{512} j \cdot \sum_{i=1}^{8} M_{ij} \tag{7.42}$$

$$mY = \sum_{i=1}^{8} i \cdot \sum_{j=1}^{512} M_{ij} \tag{7.43}$$

Both components of the parity vector can be calculated in an algorithmic way with few code lines. The obtained code is a shortened Hamming code $C[4110,4096]$ able to correct one error. The ability to recognize two errors is reached by extending the Hamming code by adding a global parity bit. In the matrix H this bit corresponds to the addition of a column of all "1" (E) and of a null row as shown in Fig. 7.21. The total parity bit can be easily obtained as the parity bit of the sum of all the rows of M. With reference to the write and read flows represented in Fig. 7.16, we have seen how to calculate the parity c_P. The syndrome s can be similarly calculated as:

$$s = r_D P + r_P \tag{7.44}$$

With this approach, the decoding of s must be performed in an algorithmic way by determining if s belongs to H and what error bit it corresponds to. Given $s = (s_E, s_X, s_Y)$, we follow these steps:

- if the syndrome s is identical to zero, there aren't errors;
- if the bit $s_E = 1$ and (s_X, s_Y) identify one bit in the matrix M (that is, $s_X \in \{1, \ldots, 8\}$ and $s_Y \in \{1, \ldots, 512\}$), only one error is detected, in position s_X, s_Y;
- if the bit $s_E = 1$ and $(s_X, s_Y) \in I$, the data matrix M is recognized as correct and one error is detected on the parity bits;
- in any other case, two or more errors are detected.

The implementation of the encoding and decoding operations requires few code lines and every byte requires two or three operations for the execution. Also with this encoding technique, it is possible to have the parity of all "1" be all "1", simply by adding $1\text{-}w$ to the parity obtained, where w is the parity of a sequence of all "1". Another simple way to satisfy this constraint is by calculating the parity of the logical inversion of M, since the parity of the null message is the codeword zero.

The code can easily fit the microcontroller structures using a word width different from the eight bits case described here.

A similar code is described by Samsung [5] for NAND applications, but it is absolutely general. The difference, compared to what was illustrated before, is in the matrix P, obtained starting from two sequences x and y sorted respectively from 0 to 511 as regards the x and from 0 to 7 as regards the y (Fig. 7.22). The matrix P is obtained through the composition of X, $X,'$ Y, Y' as represented in Fig. 7.23a. Observe that every vector in x' is the logical inversion of the corresponding x vector and that every vector y' is the logical inversion of the corresponding y vector. In this way all the constraints are satisfied so that P is a matrix for a Hamming code: in fact all the rows are unique and are equivalent neither to the vector 0 nor to the vectors of I. By construction, the code is able to detect two errors, because it can be shown that the sum of two rows of P never belongs to P and is always different from a vector of I. In fact, the sum of two vectors of P, e.g., (a, a') and (b, b'), always originates a vector of type (c,c) that belongs neither to I nor to P.

Being X, X', Y, and Y' the four matrices composing P (Fig. 7.23b), the parity vector p can be calculated as:

$$p = mP = m(Y, Y', X, X') = (mY, mY', mX, mX') \qquad (7.45)$$

Test, in order to evaluate the reliability of the solder joints at board level: that is, samples composed of the PCB and properly placed soldered DUT (devices under test) are tested.

8.2 Bend Test

Three different causes for solder joints cracking after bending a PCB can be identified. A localized bend of the PCB, possibly due to a mounting screw tying it to the external box, can cause a creep crack of the joints of the device mounted near the screw itself. Creep is the spontaneous deformation of metal materials under the effect of a constant stress at a given temperature: it means that the creep crack can occur days or even years after the product has been assembled. The second cause of cracking can be the pressure of the buttons on a mobile phone: every time a key is pressed, the PCB beneath the button tends to bend. The amount of bending, and the resulting stress on the solder joints, depends on the mechanical characteristics (elasticity, deformability, etc.) of the product. The third cause for cracking is an impact. Solder joints can crack either because of slight stress events applied cyclically or because of a major stress caused by PCB vibration; in this case too, the nature of the stress depends on the mechanical characteristics of the product and on its orientation when it is dropped.

The bend test can be used to evaluate two different but equally important features of the solder joints: mechanical endurance, and stress endurance to cyclical bending of the PCB. In the former case, the test is normally referred to as a *bending test*: it is a simple static test where PCB bending is linearly increased over time until either the joints or the components fail. In the latter case, the test is referred to as a *bending cycle test*: the PCB is cyclically bent and the mechanical endurance of the joints is measured as cycles for fatigue. Figure 8.1 shows the different setup for the two tests, which basically differ in the way both load and bending are applied to the PCB.

In the following paragraph, the procedures used to estimate, by means of the bend test, both the mechanical and the stress endurance of the solder joints are described. It is important to highlight that there are no internationally-approved standards for these tests yet; for this reason it is very difficult to compare all the experimental results that can be found in the literature.

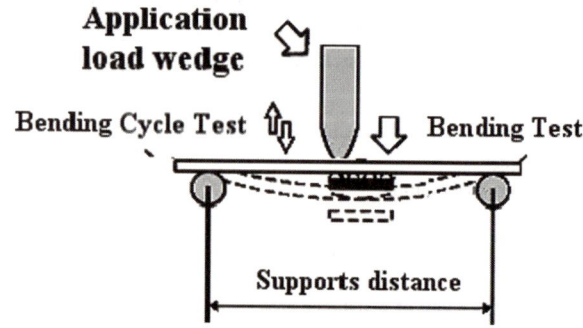

Fig. 8.1 Bend test setup

8.2.2 Bending Cycle Test

Most of the articles published on bending cycle test are based on the analysis of the effect of the change of some parameters on test results: either 3-point or 4-point configuration; test performed either in bend control (bending speed is fixed) or in load control (speed of load application is fixed); change in the distance of the supports, PCB thickness, size of the pads where solder is done, silicon die size, etc.

Table 8.1 shows a summary of the most recent studies; it is clear from the table that an international procedure for standardization of the bending test has not been set yet. All the tests shown have been performed using an Instron-type dynamometer.

8.2.2.1 Analysis of Joint Failures

Experimental results obtained using different configurations of board size and device placement show that the joints subject to cracking are usually the outermost ones in the longitudinal direction of the PCB (the bigger dimension). The first crack

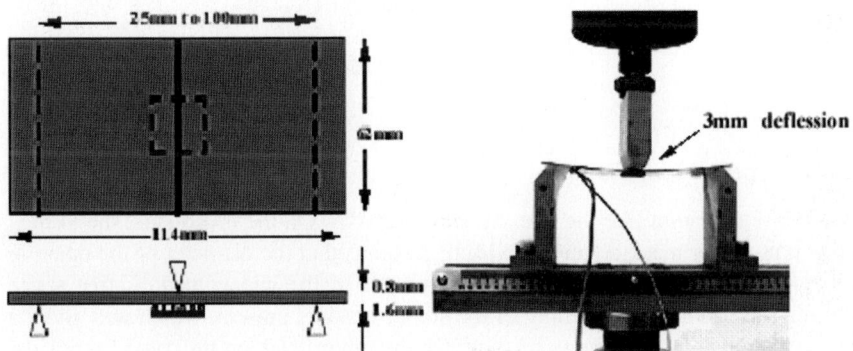

Fig. 8.5 3-point configuration (#9 in Table 8.1)

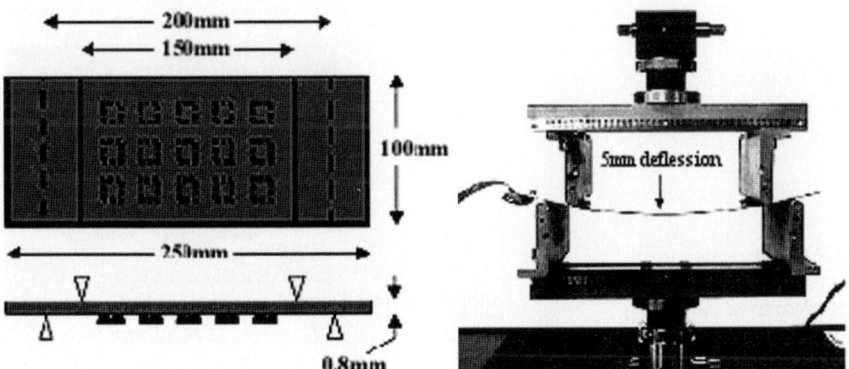

Fig. 8.6 4-point configuration (#10 in Table 8.1)

usually occurs in the corner joint, because the presence of the device causes a local change in the bending and therefore greater deformations. Because of the different bending stiffness between device and PCB, the outermost joints are subject to traction when the board bends downward, while they are subject to compression when the board bends upward.

Cracking occurs because the joints cannot stand traction and compression stresses which are rapidly cycled. The same conclusions will be true for the drop test. This is because both bend and drop tests share the same load mechanism and, therefore, the same failure pattern: joint breaking is mainly caused by PCB bend.

Fig. 8.7 PCB used for configuration #12 in Table 8.1; PCB size is 115 mm × 77 mm × 1 mm; devices are BGA 7 × 7 mm with 84 balls

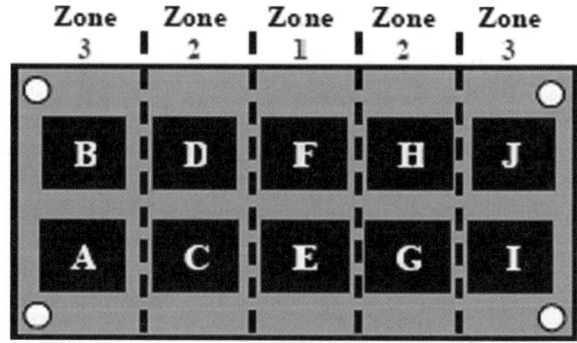

8.2.2.2 Types of Joint Failures

Let's consider a joint like the one shown in Fig. 8.8. On the board side, the joint is called non-solder mask defined (NSMD); it means that the diameter of the opening of the *solder mask*[2] is larger than the diameter of the copper pad. A free space between the pad and the opening of the solder mask is present, which will allow a better adherence of the joint to the pad. On the other hand, on the component side, the joint is SMD (solder mask defined): the opening of the solder mask is smaller

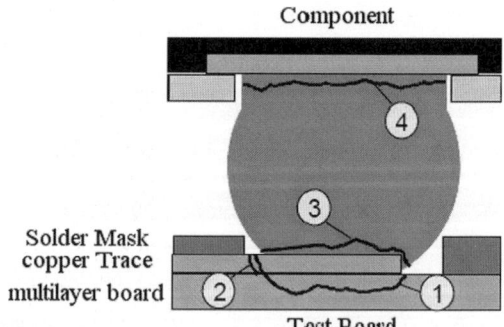

Fig. 8.8 Schematic representation of the failure modes observed during bend test

___

[2] Layer of passivation material which covers the outermost conductive layer (copper trace).

than that of the copper pad. In this kind of joint, 4 different types of crack can be observed during a bend test.

The first failure mode is the delamination of the pad on the board. Delamination is the process of "exfoliation" which causes the detaching of the pad from the board. Once detached, the pad can freely move either upwards or downwards, depending on the board bending direction. Deterioration of this "mechanical support" can cause a stress-induced crack of the copper trace on the board. The second failure mode is the cracking of the copper trace without a corresponding crack in either the solder mask, the pad, or the laminate. In general, the cracking in the copper trace starts in the area next to the solder mask opening. These first two types of failures can typically be observed in all those experiments characterized by more severe stress conditions (greater board bend).

The third failure mode is characterized by a crack induced by joint stress near the pad on the board. The crack usually starts on the perimeter of the joint, where it does not cover the pad laterally; this is the weakest point, because the joint here is really defined by the solder mask, and therefore the stress concentrates here. Figures 8.9 and 8.10 show two photographs depicting this kind of failure: one shows an analysis performed using the so-called dry and pry technique, the other shows a micro cross-section of the joint. Dry and pry is a technique based on the use of a colouring fluid in which the device under test is dipped. After drying it, by mechanically detaching the part from the board it is easy to observe the progress of the cracking in the joint: the fluid colours all those parts which were already cracked before detaching the part.

The fourth failure mode is the joint cracking induced by stress at the interface with the device. Figure 8.11 shows the analysis of this kind of failure performed by means of dry and pry, while Fig. 8.12 shows a micro cross-section of the joint. In general, the first two failure modes can be observed in association with high stress conditions, while the third and fourth modes occur in less severe stress conditions.

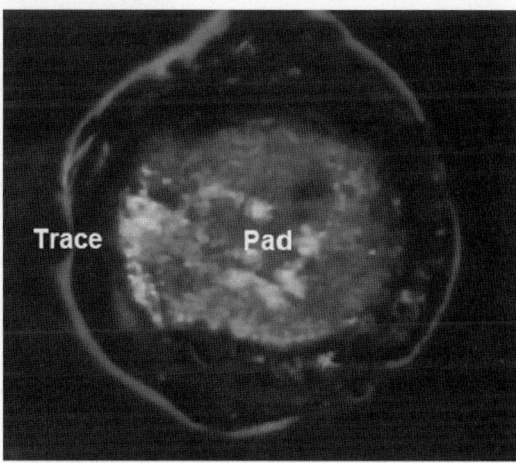

Fig. 8.9 Failure mode #3: dye penetrant colours the cracked surface of the joint near the interface with the copper pad of the PCB

Fig. 8.10 Failure mode #3: micro cross-section of the cracked joint

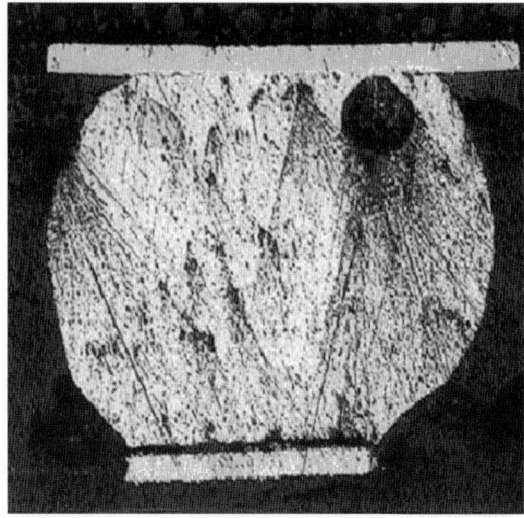

Fig. 8.11 Failure mode #4: dry and pry analysis

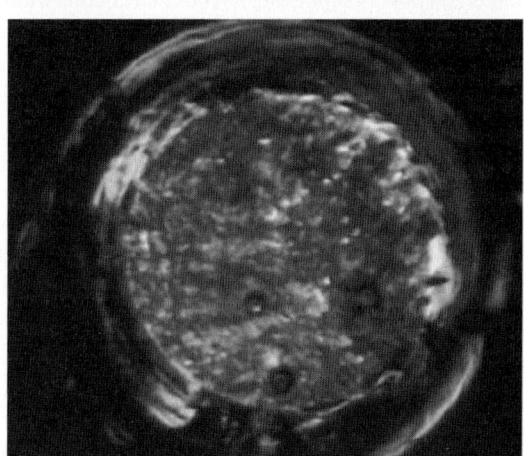

Fig. 8.12 Failure mode #4: micro cross-section of the cracked joint

8.2.2.3 Effect of Distance Variation of the Standoffs

Analyzing experimental results reported by previous publications, it is evident that a decreased distance between the standoffs causes a reduction in time to failure. This is due to the fact that the load acting on the PCB is inversely proportional to the cube of the distance between the standoffs, for a given value of deformation.

It is also demonstrated that a cracked joint failure mode turns into a crack into the copper trace of the board as deformation, and therefore the applied level of stress, increases.

8.2.2.4 Effect of PCB Thickness

Increasing experience in the field of reliability testing has led to the conclusion that the stress on the joints can be reduced by increasing both the thickness and the stiffness of the board the devices are mounted on. This is only partially true; in fact, the analysis of both the experimental and the simulation results of the bend test performed on BGA devices has shown that the percentage of failed joints considerably decreases both for high and for low stiffness boards. This contradictory conclusion can be easily understood, considering the way the test is performed. In case of load control, the use of a thicker board leads to lower levels of bending and deformation of the board plane; therefore, since both deformation and the load acting on the joints are proportional to the deformation on the plane of the board, the joints are less stressed. On the other hand, in the case of bend control, a joint on a thicker board must be able to sustain a higher level of stress than a joint on a thin board. This is because both deformation on the board plane and the load acting on the joints increase as the board gets thicker, given a bending value; in particular, the load is proportional to the cube of board thickness. Therefore design optimization of the PCB must take into account these two contrasting factors, together with the final application of the circuit.

8.2.2.5 Effect of Pad Size

The literature reports several studies performed on BGAs which have shown how time to failure highly depends on the size of the pads, both on the board and on the device itself. In general, if the size of the pad is increased on the side of the joint where the failure occurs more often, reliability is improved. A bigger pad causes an increase in both the resisting area and the length the crack has to be in order to cause a failure. On the other hand, if the size of the pad is increased in the area opposite to the one where the joint tends to crack, a stress concentration occurs, which causes reduced reliability of the joint itself. Simulations performed to analyze the effect of pad size in bend tests has shown that the size of the pad on the component side has a greater influence on time to failure than the size on the other side of the board.

8.2.3 Load Control Versus Bend Control

As already mentioned in previous sections, it is possible to perform the bend test in two ways: load control and bend control. By using the bend control method, it has been proven that the time to failure of the devices under test decreases when:

- distance between standoffs decreases;
- PCB thickness increases;
- package thickness increases.

This is due to the fact that, for a given bend, reaction force increases proportionally to any of the parameters just mentioned. In the case of load control, a cyclic constant load is applied to the device under test and the PCB is bent according to the elastic characteristics which characterize the set of PCB + device(s). In the case of load control, the behaviour is the opposite than in the case of bend control; for instance, the thicker the PCB is, the better is the time to failure of the devices. In general, the bend control test is preferred, because the design and manufacturing of test equipment for this kind of test is easier.

8.2.4 Conclusions

In previous sections, the bend test has been presented; it is a recently introduced test used to evaluate the reliability of solder joints for BGA devices mounted on a PCB. As reported, it is possible to execute this test using two methodologies which differ both in the way the load is applied and the set of properties of the joints that can be investigated:

- bending test, useful to evaluate mechanical resistance of the joints; during the test, either PCB bending or the applied load increases over time according to a given law, until a failure in the joints occurs;
- bending cycle test, useful to evaluate stress resistance of the joints when the PCB receives a cyclical bending stress; stress resistance of the joints is measured in cycles to failure (in other words, the number of cycles after which a failure occurs is recorded).

Both tests can be carried out using either a load or a bend control, implementing either a 3-point or a 4-point configuration. Neither bending test nor bending cycle test is regulated by internationally acknowledged standards; only internal specifications from either manufacturers or customers exist, and therefore it is hard to compare all the results reported in the literature. It has been proven that both bending cycle tests and drop tests are characterized by the same mechanisms of load and failure: breaking of the joints is mainly caused by PCB bending and the main stress is caused by traction. This mechanism is completely different from the one which characterizes thermal cycles, in which the failure of the joints is caused by the different coefficient of thermal expansion (CTE) of the materials which constitute the package, and the main stress is a cutting stress. In the case of the bending cycle test, the devices mounted in the centre of the PCB are most critical, and for these devices the most critical and break-prone joints are those in the corners. Four different failure modes for the bending cycle test have been identified. When high stress conditions are present, the crack tends to originate from (and propagate inside) the copper trace of the board rather than in the joint. In the case of the bending cycle test executed in bend control, the time to failure of the joints decreases, both decreasing the distance between the standoffs and increasing the thickness of the board. In the case of the bending cycle test executed in load control, the opposite behaviour occurs.

8.3 Drop Test

In the literature, it is possible to find publications which confirm and demonstrate how mechanical shocks can cause a cyclical bending of the board—a sort of oscillation whose amplitude decreases over time, which causes a cracking in the joints from mechanical stress. Such a cracking is defined as "intermittent," because it corresponds to cyclical changes and sudden increases of the dynamic electrical resistance in the joints themselves. Therefore, both PCB design and mounting conditions are very important, because they determine the dynamic characteristics of the sample, i.e., the way the whole set of board and device(s) reacts and moves according to the specific mechanical stress applied to it. Unlike with the bend test, an international JEDEC (Joint Electronic Device Engineering Council) standard is available for the drop test, which defines the standard size, structure, and characteristics for the materials used for the PCB, so that the performance of different devices can be compared. In this section, the method of the JEDEC standard is briefly described; and then the test mechanics, the dynamic responses of the PCB and the joints, and the criteria and analyses of the failures are taken into account. In the final part, the effect of the variation of some parameters (size and geometry of the PCB, materials used to create the joints, number of constraints imposed to the board) on test results is investigated.

8.3.1 JEDEC Standard

A drop test is classified according to the behaviour of *acceleration* or *pulse*, which is in turn characterized by two parameters:

Peak acceleration: the maximum value for the acceleration caused by the impact of the device on a surface, which can be calculated using the equations of the dynamics (energy conservation and pulse principle) considering the acceleration waveform (i.e. the behaviour over time).

Pulse duration: the time interval between the moment when acceleration reaches 10% of its peak value (during the ascending phase) and the moment when acceleration gets back to 10% of its peak value (during the descending phase).

8.3.1.1 Test Equipment

The equipment for the drop test must be able to cause mechanical shocks up to a maximum acceleration value of 2900 g (i.e., 2900 times the acceleration of gravity), with a pulse duration between 0.3 and 2 milliseconds, with an absolute tolerance of ± 30% and a speed change between 122 and 543 cm/s. Speed change is defined

Table 8.2 Drop test levels

Test condition	Height of fall (inches)/(cm)	Speed change (in/s)/(cm/s)	Acceleration peak (G)	Pulse duration (ms)
H	59/150	214/543	2900	0.3
G	51/130	199/505	2000	0.4
B	44/112	184/467	1500	0.5
F	30/76.2	152/386	900	0.7
A	20/50.8	124/316	500	1
E	13/33.0	100/254	340	1.2
D	7/17.8	73.6/187	200	1.5
C	3/7.62	48.1/122	100	2

as the arithmetical sum of the speed of the sample before and after the impact (speed at the time of bouncing). Acceleration must have a sinusoidal behaviour over time, with a maximum distortion of $\pm 20\%$ with respect to the specific peak value. Table 8.2 shows all the levels of drop test specified and accounted for in the JEDEC standard.

8.3.1.2 Board Design

The board used for the test is multilayer, symmetrical with respect to the middle layer, manufactured according to a type $1 + 6 + 1$ configuration. Nominal thickness of the board must be 1 mm. Table 8.3 summarizes the thickness, the percentage of area covered by copper, and the type of material for each layer.

The surface finish of the PCB is composed of an OSP (organic solderability preservatives) coating, in order to prevent an oxidation of the copper pads before surface mounting of the devices. The component must be tested by means of both microvia pads and no-microvia pads, which can be done if a double-sided board is designed. On side A of the board, microvia pads are present, while no-microvia pads are on side B. For the sake of the test, components must be mounted on one side at a time. Pad sizes are shown in Table 8.4. All the pads must be NSMD (non-solder mask defined), with a difference of 0.15 mm between the diameter of the pad and the opening of the solder mask.

The thickness of the traces, including the traces which connect both internal layers of the board and the joints, must be equal to 75 microns inside the area where the components are located. Outside that area, the thickness of the traces must be greater than or equal to 100 microns.

The board must implement a *daisy chain* scheme: once the devices are mounted on the board, a sort of network is created among the interconnections of the joints of the components; a single chain is usually created among all the joints for each device. Each device is therefore assigned a unique output channel so that a crack can be easily spotted. Where needed, it is possible to add test points inside a single network in order to ease the detection of the failing joint.

Table 8.3 Design guidelines for the drop test board

Board layer	Thickness (microns)	% of Area covered by copper	Material
Solder Mask	20		LPI
Layer 1	35	Pad + Traces	Copper
Dielectric 1–2	65		RCC
Layer 2	35	40%, including links between daisy chains	Copper
Dielectric 2–3	130		FR4
Layer 3	18	70%	Copper
Dielectric 3–4	130		FR4
Layer 4	18	70%	Copper
Dielectric 4–5	130		FR4
Layer 5	18	70%	Copper
Dielectric 5–6	130		FR4
Layer 6	18	70%	Copper
Dielectric 6–7	130		FR4
Layer 7	35	40%	Copper
Dielectric 7–8	65		RCC
Layer 8	35	Pad + Traces + links between daisy chains	Copper
Solder Mask	20		LPI

Table 8.4 Recommended values for pad diameter and opening of the solder mask on the PCB

Component I/O pitch (mm)	Pad diameter on PCB (mm)	Solder mask opening (mm)
0.5	0.28	0.43
0.65	0.3	0.45
0.75/0.80	0.35	0.5
1	0.45	0.6

8.3.1.3 Board Layout

The layout of the board under test is schematically shown in Fig. 8.13. Four holes must be drilled at the corners of the board, in order to allow a proper placement for the test.

All the components must be placed inside a 95 mm × 61 mm rectangle defined by the outermost sides of the outermost components. Proper connection points must be designed for the cables, one for each device (each cable is connected to a single device).

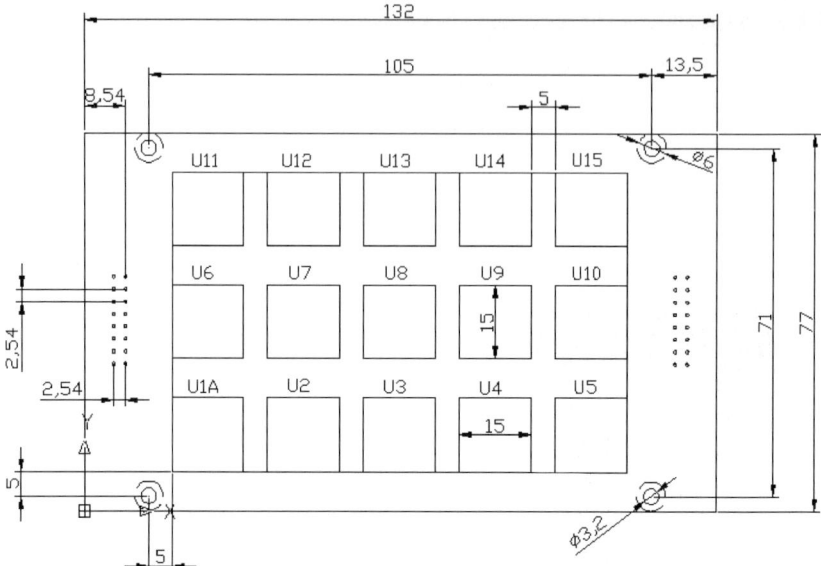

Fig. 8.13 Geometry and size of the board, according to JEDEC standard

8.3.1.4 Number of Components and Samples To Be Tested

The board shown in Fig. 8.13 can host up to 15 components.

Considering that the board is constrained in its four corners, the locations occupied by the devices include the worst case, where the maximum bend of the board occurs (location U8), the case where the proximity of the constraints is of influence (locations U1, U5, U11, and U15), and all the intermediate cases, thus allowing a lot of information to be obtained from the test. It is possible to mount on the PCB a different number of devices. For instance, options for either 1 or 5 devices are considered, and devices should be placed as follows:

- for 1-component configuration: location U8;
- for 5-component configuration: locations U2, U4, U12, U14, and U8.

Depending on the number of mounted devices, Table 8.5 can be used to determine the minimum number of samples to be tested and the corresponding number of devices, in order to get both sufficient and reliable results. In case of rectangular devices, they should be assembled in such a way that their longest side is parallel to the longest side of the board.

8.3.1.5 Procedure

Equipment Used

Figure 8.14 schematically shows the typical equipment used for a drop test.

Table 8.5 Values and number of boards and devices required by the test

Number of components per board	Number of boards		Total number of components
	Side A (Via in pad)	Side B (No Via in pad)	
15	4	4	120
5	4	4	40
1	10	10	20

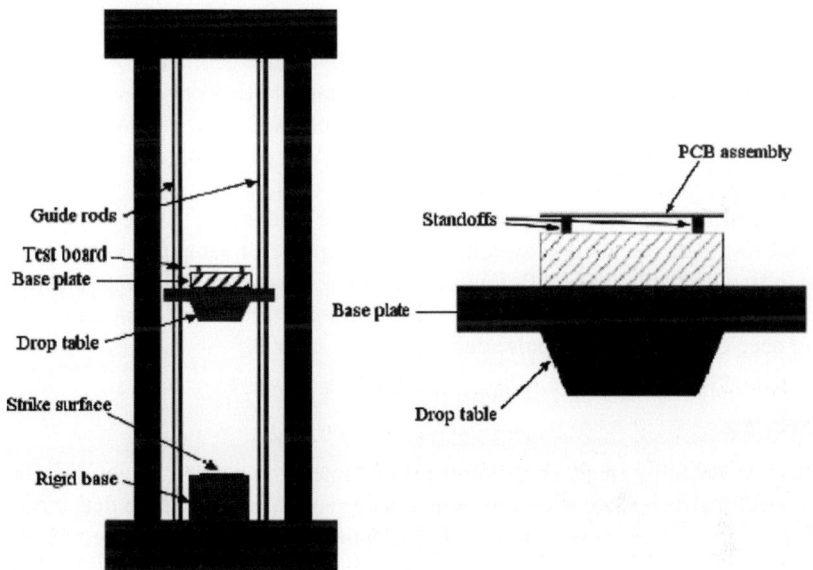

Fig. 8.14 Typical equipment used for drop test

The *drop table*, whose size is $210 \times 210 \times 70$ mm, is dropped from a given height, subject to the force of gravity only, along two cylindrical guide rods in order to cause an impact on a rigid base. Such a base is coated with a proper material (*strike material*), whose characteristics allow obtaining both the desired level of acceleration (measured as a multiple of the acceleration of gravity) and the desired pulse duration.

On the drop table, a *base plate* ($140 \times 76 \times 15$ mm) is rigidly mounted, equipped with specific *standoffs* (6 mm external diameter, 3.2 mm internal diameter, 10mm height) used to secure the PCB under test, as shown in Fig. 8.15.

The board is positioned on the tester horizontally by means of four screws in its corners, components facing down; that is the configuration which causes the maximum bend of the board (worst case). The pulse which must be applied to the sample under test corresponds to JEDEC condition B (1500 g, 0.5 milliseconds, sinusoidal pulse).

Fig. 8.15 Detail of the base
plate

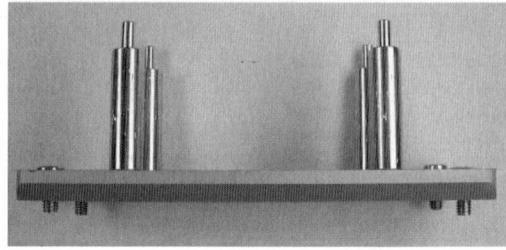

Test Characterization

An acceleration sensor is placed in the centre of the base plate in order to monitor the
pulse transmitted to the board at the moment of the impact measured as acceleration
versus time. In specific cases, a *strain gauge*[3] can be used, which is mounted on
the board in correspondence with the central component on the opposite side with
respect to the devices, as shown in Fig. 8.16.

Both measurement systems (acceleration sensor and strain gauge) must be
connected to a data acquisition system which is capable of sampling values at a
minimum frequency of 20 KHz. Before starting the test, both the fall height and the
strike material must be properly chosen in order to get the desired conditions (in
terms of acceleration peak and pulse duration).

Test Execution

The test is executed by freely dropping the drop table from a given height. During
the test, electrical resistance must be continuously monitored in order to detect any
potential failure immediately. The board is dropped either for a maximum of 30

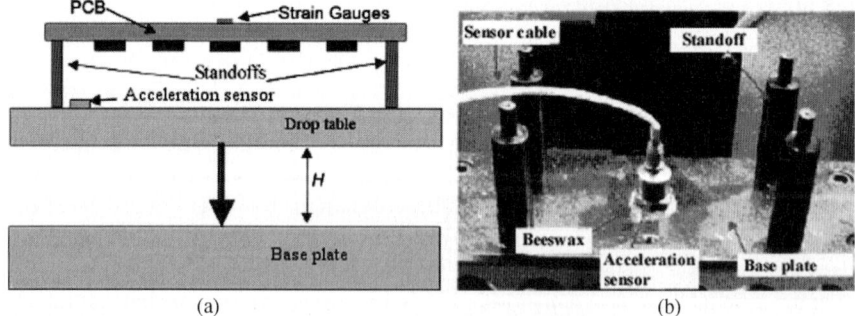

Fig. 8.16 (**a**) Schematic representation of the setup used to characterize the test; (**b**) Position of the
acceleration sensor used to measure the acceleration impressed to the board as a function of time

---

[3] Instrument used to measure the bend (along both the longitudinal and the cross directions) produced on the board by the impact.

times or until 80% of failures occur. The pulse characterizing the impact must be periodically checked in order to ensure it stays within the specified limits.

8.3.1.6 Failure Criteria and Analysis

Failure can be defined in two different ways, depending on which system is used to monitor electrical resistance:

- When an *event detector* is used, a failure is defined as the first case of resistance discontinuity whose value is greater than 1000 ohms for a minimum duration of 1 millisecond, followed by three similar events in the next 5 drops.
- When *high speed data acquisition systems* are used, a failure is defined as the first occurrence of either a resistance value equal to 100 ohms or an increase of 20% with respect to the initial resistance value if this is greater than 85 Ohm, followed by three similar events in the next 5 drops.

A visible partial detaching of a component is considered a failure as well, even if a corresponding increase in resistance does not occur. For each test lot, a group of 5 devices must be analyzed in order to understand the root cause which led to the cracking and the failure mechanism. Devices to be analyzed must be selected in such a way that they occupy different positions on the board. Because of both their symmetrical placement and the securing mechanism used for the board, it is possible to combine the 15 components into 6 groups: all the devices belonging to the same group will be stresses in the same way, and therefore they will experience the same failure mode (see Table 8.6). Groups E and F can be combined together because the bend of the board in correspondence with those components can be considered very similar.

8.3.2 The Mechanics of the Drop Test

All the impact processes, the drop test included, involve a series of energy transformations. Two main forms of mechanical energy exist: *potential* energy and *kinetic*

Table 8.6 Devices mounted on the board grouped according to the stress they are submitted to

Group	Number of components within the group	Position of components on the board
A	4	U1, U5, U11, U15
B	4	U2, U4, U12, U14
C	2	U6, U10
D	2	U7, U9
E	2	U3, U13
F	1	U8

energy. Potential energy is the energy belonging to a body due to its position inside the earth's gravitational field, and it depends on the height of the body from the ground. Kinetic energy, on the other hand, represents the energy that a body has because of its motion, whose intensity is proportional to the square of the speed of the body itself. As already pointed out, the drop table is dropped freely from a given height, therefore its potential energy gets gradually converted into kinetic energy. In consequence of the impact of the drop table against the rigid base, a force (pulse) develops, which is characterized by a certain level of g (unit of measure which indicates multiples of the acceleration of gravity) and by a pulse duration. In fact, the force is measured by means of the acceleration produced by the impact, monitored by an accelerometer, whose pattern over time is represented by a sinusoidal function, according to JEDEC specifications (therefore characterized by a peak value and duration). Impact forces are transmitted to the solder joints and lastly to the components by the mounting screws. At the same time, at the moment of the impact, the force of inertia causes the bend of the board, which induces a stress state in the joints.

From a dynamic point of view, the initial potential energy of the sample under test, K, can be derived from:

$$K = mgH, \tag{8.1}$$

where m is the mass of the sample (board + devices), g is the acceleration of gravity ($9.81 \ m/s^2$) and H is the fall height.

By applying the law of preservation of energy and neglecting any potential loss due to friction forces, potential energy is totally converted into kinetic energy, T, calculated as:

$$T = \frac{1}{2}mV_b^2. \tag{8.2}$$

By combining equations (8.1) and (8.2) it is possible to derive the speed, V_b, at the moment immediately before the impact as a function of the fall height H:

$$V_b = \sqrt{2gH}. \tag{8.3}$$

By applying both pulse theory and momentum theory, it follows that the value of the speed after the impact, V_a, falls in the range between 0 (in the case of no rebound) and $-V_b$ (in the case of complete rebound). Letting V_a be a fraction of V_b, $V_a = cV_b$:

$$-mcV_b - mV_b = -\int_0^T mG(t)dt, \tag{8.4}$$

$$V_b = \frac{1}{1+c} \int_0^T A(t)dt,$$ (8.5)

where c represents the elastic coefficient which varies from 0 (perfectly plastic impact, $V_a = 0$) and 1 (perfectly elastic impact, $V_a = V_b$), $A(t)$ represents the value of acceleration during the impact at time t, and T represents pulse duration. The JEDEC standard introduces the so-called rebound coefficient C:

$$C = 1 + c.$$ (8.6)

This coefficient is set equal to 1 in the case of no rebound, and it is set equal to 2 in the case of complete rebound. As specified by JEDEC, the acceleration over time, during the impact, must be of sinusoidal type. The curve can therefore be approximated by the equation:

$$A(t) = A_0 \sin \frac{t \cdot \pi}{T},$$ (8.7)

where A_0 represents the acceleration peak or maximum shock level. Such a curve, shown in Fig. 8.17, is just a theoretical representation of the real behaviour of acceleration as a function of time. The real acceleration curve derived during the test is not a perfect sinusoidal function (see Fig. 8.18), but it tends to a triangular shape. The area beneath the curve represents the energy which is absorbed by the board during the impact, and then transferred to both joints and devices. By using equation (8.7), equation (8.5) becomes:

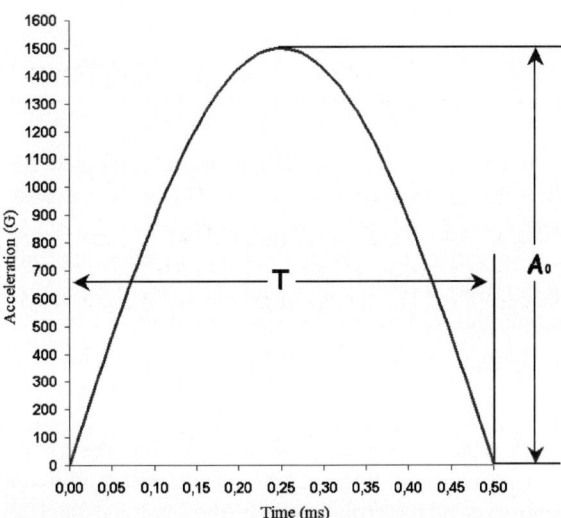

Fig. 8.17 Theoretical representation of the sinusoidal behaviour of the acceleration as a function of time

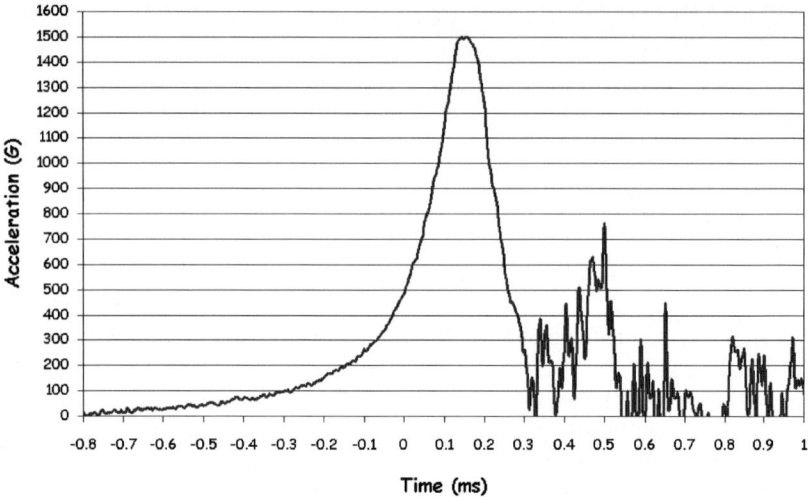

Fig. 8.18 Real shape of acceleration as a function of time during the test

$$V_b = \frac{1}{1+c} \int_0^T A_0 \sin \frac{\pi t}{T} dt. \tag{8.8}$$

Considering the case of perfectly plastic impact (no rebound $\Rightarrow c = 0$), it is possible to define the acceleration peak A_0 as a pure function of fall height by solving the integral and by using equation (8.3) to calculate V_b:

$$V_b = \sqrt{2gH} = 2A_0 \frac{T}{\pi}, \tag{8.9}$$

$$A_0 = \frac{\pi \sqrt{gH}}{T \sqrt{2}}. \tag{8.10}$$

In the case of perfectly elastic impact (complete rebound $\Rightarrow c = 1$), A_0 is twice the value given in equation (8.10). Therefore it is possible to use equation (8.10) to determine the fall height required to get the desired level of G (acceleration peak). It is important to highlight that this equation does not take into account the effect of the strike surface, which should be determined experimentally. By setting the fall height, the $A_0 T$ product remains constant,

$$A_{01} T_1 = A_{02} T_2, \tag{8.11}$$

because the transmitted pulse, and therefore the energy absorbed during the impact, remains constant. Before starting the test, it is advisable to determine both the characteristics of the strike surface (the kind and thickness of the material chosen to cover it) and the fall height required to get the desired shape for the acceleration curve (in terms of both A_0 and T). It should be remembered that if the thickness of the material covering the surface is increased, the peak acceleration value decreases and duration increases.

8.3.3 The Dynamics of the Drop Test

Since electronic components are mounted on a board, the behaviour of the board itself during the impact directly influences their performance. At the moment of the impact, the pulse is transmitted to the board in a very short time; at the same time, the speed of the drop table decreases to zero, while the board together with its components is still moving downwards. Therefore the PCB tends to bend first downwards because of the force of inertia, and then upwards because of the forces of elastic recall (Fig. 8.19). Afterwards, it will cyclically bend downwards and upwards until all the energy absorbed in the impact is dissipated. Figure 8.20 represents the behaviour of board deformation along the longitudinal direction measured on six samples exposed to a drop test with a pulse duration of 1 millisecond. The board repeatedly undergoes traction and compression. Previous publications suggest that output acceleration (the acceleration the board is subjected to after the impact) is closely related to the deformation pattern along the longitudinal direction.

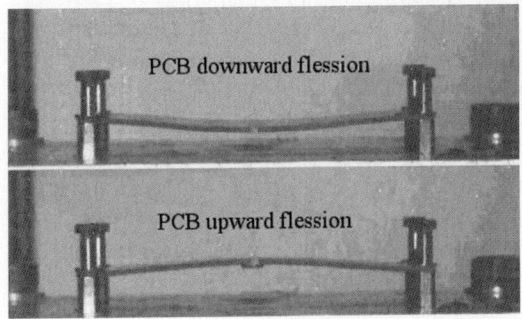

Fig. 8.19 Vibration of the board at the time of the impact

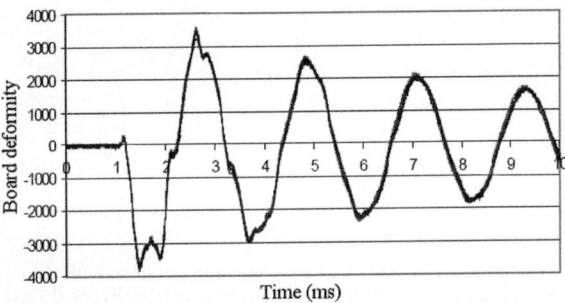

Fig. 8.20 Behaviour over time of the dynamic deformation of the board along longitudinal direction

Fig. 8.21 Bending of the board due to impact; both deformation and output acceleration vary cyclically with the same frequency

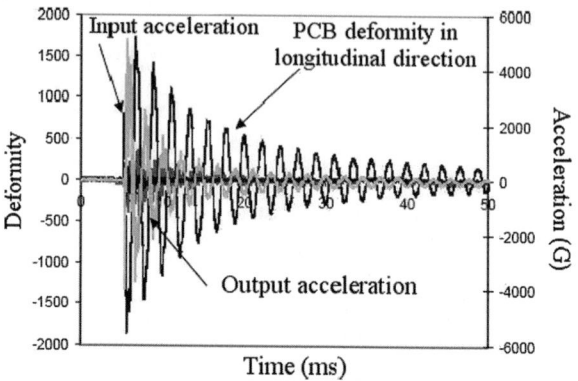

When the board bends downward, the opposite side experiences compression along the longitudinal direction. The maximum value of deformation occurs when the maximum bend is reached. At that moment, the speed in the centre of the board is zero, which means that when the board bends downwards, acceleration is at its maximum in the opposite direction.

Therefore, in the case of negative deformations, acceleration is positive (i.e., it is going upwards), and vice versa, which confirms that output acceleration has the same cyclic frequency which characterizes board bend. Figure 8.21 represents the behaviour over time of the deformation of the board, which cyclically bends until it gets to around ten cycles per drop, and the amplitude of which gradually decreases over time. Therefore, on top of the mechanical shock, the effect of the stress caused by vibrations having variable amplitude should be taken into account in order to evaluate the reliability of solder joints properly.

Figure 8.22 schematically shows the effect of the bend of the board on the solder joints for a device placed in the centre of the PCB. Because of the different bending stiffnesses of the board and the device, the outermost corner joints will be subjected to traction when the board bends downwards and to compression when it bends upwards. Since this kind of joint cannot tolerate a rapid series of small traction and compression stresses which occur in a very short time, a crack of the outermost joints can be expected.

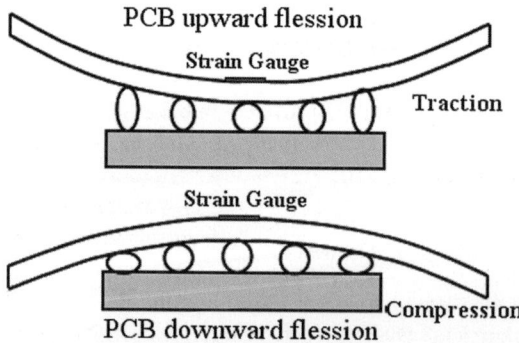

Fig. 8.22 Effect of bending of the board on the solder joints

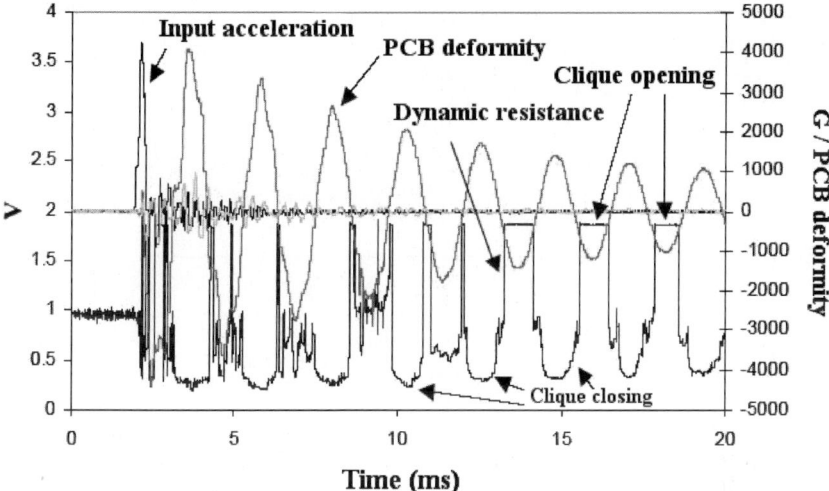

Fig. 8.23 Cyclical behaviour of dynamic electrical resitance of the joints during the test

Once the crack starts, it will propagate when the joint is subjected to traction (board bending downwards), while it will tend to close when the joint is in compression (board bending upwards). This fact represents another experimental proof of how the reliability of the solder joint depends heavily on the dynamic response of the board, and of how both cyclical bending and vibration are the main causes of failure of the joints due to impacts. During the test, electrical resistance is another parameter which tends to vary cyclically after a number of drops, as shown in Fig. 8.23; it is worth recalling that electrical resistance is inversely proportional to the total contact surface (the sum of the contact surfaces of all the joints in the device). Resistance will reach its maximum value ($R \rightarrow \infty$) when the board bends downwards, i.e., when the joint is in traction and the crack, if present, tends to propagate. Resistance value drastically drops when the board bends upwards; joints are in compression and the contact is restored.

8.3.4 Failure Criteria

Studies reported in previous publications have shown that stresses acting on the joints vary cyclically over time with the same frequency as dynamic deformations of the board. Figure 8.24 shows variation over time of the stress acting on the critical joint (the outermost corner joint); S_x represents the stress acting along the longitudinal direction of the board, S_1 is the main stress, S_z is the stress acting orthogonally to the board (also known as peeling stress), S_{zz} represents the edge stress, and $S_{eq.v}$ is the equivalent stress calculated using the resistance criteria of Von Mises (given a tensional state induced by one or more stresses acting in different, but orthogonal,

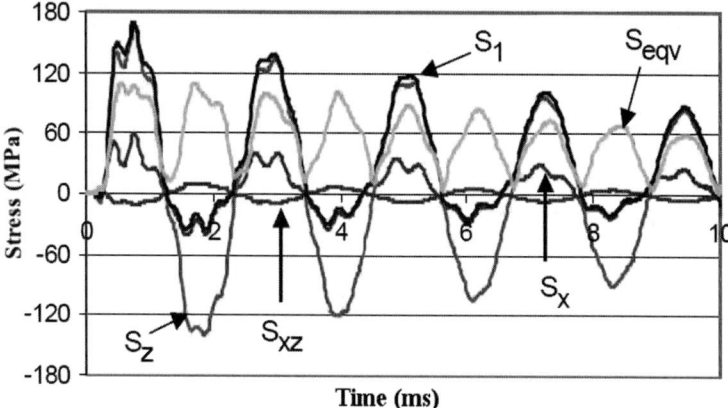

Fig. 8.24 Behaviour of the stresses acting on the joint during the test as a function of time

directions, it is possible to determine the value of an equivalent resulting stress, which represents the same tensional state).

Peeling stress is the most critical one during the test, and therefore it can be used as a failure criterion for design purposes, i.e., to predict potential failures. Since such a stress is mainly caused by board vibration, it can be inferred that the cyclical bend of the PCB during the test represents the main failure mechanism, especially for the devices placed in the central area of the board, in case the sample (PCB + devices) is secured by means of 4 points.

Figure 8.25 compares the behaviour over time of the input acceleration (i.e., the acceleration characterizing the pulse, having sinusoidal shape and classified on both its peak level and pulse duration), of board bend along the longitudinal direction, of dynamic electrical resistance of the joints, and of peeling stress at the most critical points.

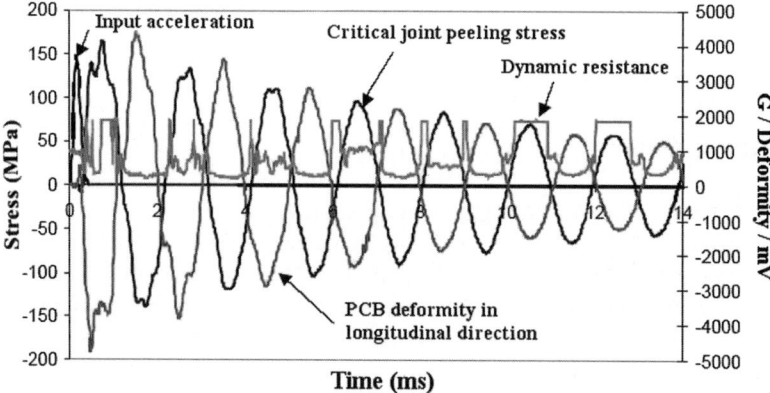

Fig. 8.25 Dynamic response of the sample during the test

Since the strain gauge is mounted on the opposite side of the board with respect to the components, peeling stress is positive when deformation is negative, and therefore when the joint is in traction; then the crack tends to propagate and dynamic resistance is at its maximum value. Vice versa, when the negative peak of deformation is reached, peeling stress is negative, the joint is compressed, the crack tends to close, restoring the contact and thus considerably decreasing resistance value. The coincidence of cyclic behaviours of stress, bending, and electrical resistance is another proof of the fact that start, propagation, and finally failure of the crack are mainly caused by the vibration the board is subjected to because of the impact.

8.3.5 Failure Analysis

Analysis of the potential failure caused by the drop test will be done considering a board secured with 4 screws positioned in its corners. Figure 8.26 describes a drop test characterized by a peak acceleration of 1500 g and a duration of 0.5 milliseconds, and it is possible to see that the maximum value of peeling stress does not occur at the same time as acceleration reaches its peak value, but at a later time instead, exactly when the board is subjected to maximum bending caused by forces of inertia after the impact.

Joint cracking usually occurs at the interface and it is caused by a combination of mechanical shock and board bend. Figure 8.27 shows how deformation of the board is distributed along the longitudinal direction in the case of maximum bending both downwards and upwards.

In both cases, the maximum deformation occurs in the centre of the board (darker colour); the bending value tends to decrease moving away from the centre (colours get brighter) until standoffs are reached, when bending gets reversed (from compression to traction and vice versa). In this specific case, the board is deformed more along the longitudinal direction; for this reason, the distance of the devices

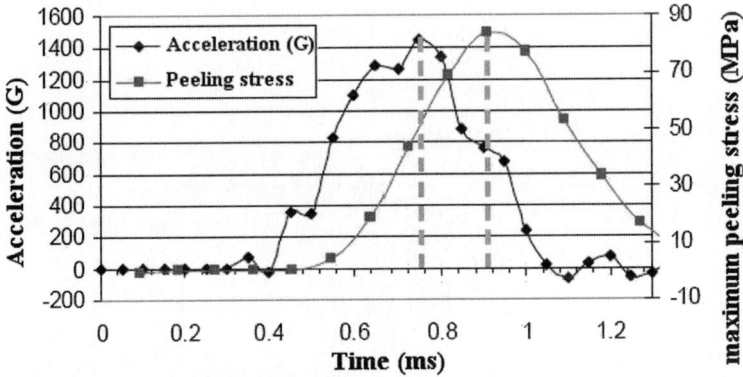

Fig. 8.26 Behaviour of input acceleration and orthogonal stress on the joint (peeling stress) during the impact for a drop test characterized by a peak of 1500 g and a duration of 0.5 ms

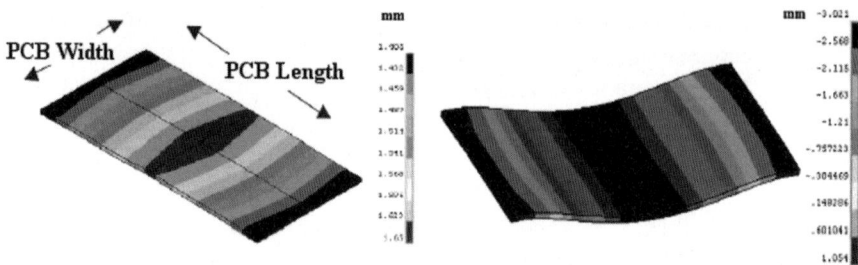

Fig. 8.27 Distribution of board deformation during maximum bending, both upwards and downwards

from the centre of the board becomes of the utmost importance, since the centre is the most critical area, where bending is at its maximum.

Considering the distance from the centre, it is possible to partition the board into five zones; devices placed in the same zone will be subjected to the same stresses and therefore will experience the same failure mechanisms. The farther the device is from the centre of the board, the less it is subjected to failure, and therefore the less it will be able to tolerate a higher number of drops. Within the same device, the most critical joints are the outermost along the longitudinal direction. The first failure usually occurs in the corner joint, which has to cope with a greater deformation and therefore a higher level of stress, as shown in Fig. 8.28.

A higher degree of stress can be measured at the joint/copper pad interface both on the board and on the component side. Which one tends to crack first depends on their adhesion resistance. Most of the experiments have shown that the critical interface is on the component side, and that the crack starts and propagates in the region where intermetal compounds (a byproduct of the soldering process) are present.

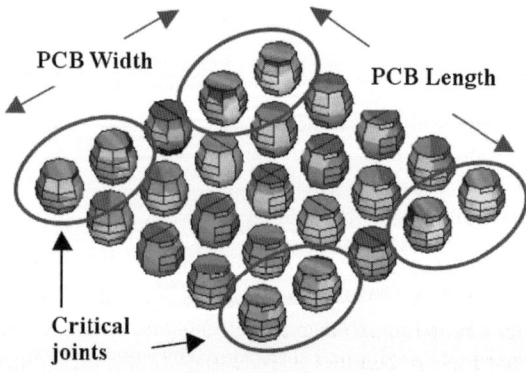

Fig. 8.28 Distribution of stresses in the joints in conjunction with maximum board bend upwards (maximum peeling stress 128 MPa)

8.3.5.1 Failure Process of the Joints

Considering and monitoring the behaviour over time of dynamic resistance, it is possible to analyze in a more detailed way the failure process of the joints. Dynamic resistance means the electrical resistance of the joints evaluated "dynamically," i.e., while the test is being executed. Static resistance means the electrical resistance of a device evaluated "statically," i.e., stopping the test. Considering these parameters and above all the value that dynamic resistance has over time, the failure process of the joints can be divided into three main phases (see Fig. 8.29): start of the crack (phase 1), propagation of the crack (phase 2), and opening of the crack (phase 3). In the first phase, both static and dynamic resistance gradually grow with the number of falls. This implies that a crack has already started and that the cross-section of the joint is becoming smaller and smaller. In the second phase, some peaks begin to occur in the graph of dynamic resistance (number of falls equal to N_1); the crack is propagating, the area of the cross-section is reduced further, and static resistance increases. After following falls, the amplitude of the peak of dynamic resistance increases more and more until the so-called "intermittent crack" $(R \rightarrow \infty)$ occurs, in correspondence with a number of falls equal to N_2. From this moment on, the crack crosses the whole interface with the joint and an electrical discontinuity (whose duration increases over and over) occurs.

In the third phase the process goes on, the gap between the joint and the copper pad gets larger, and resistance tends to infinity for most of the time. When the value of both the static and the dynamic resistance becomes virtually infinite permanently, the joint is considered to be irreparably broken. Unfortunately, the resistance monitoring systems usually available are not able to provide such a detailed measurement of the resistance, but only the data related to the permanent crack of the joint.

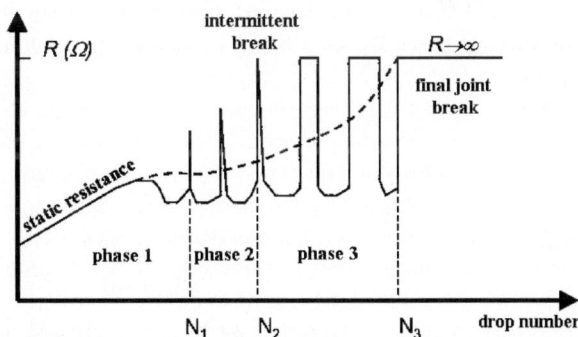

Fig. 8.29 Failure process of the joints

8.3.5.2 JEDEC Board

Considering Fig. 8.13, and in accordance with the conclusions and the results mentioned above, the most critical placements for the devices are U8, U13, and U3 (mentioned in growing order of severity), since they are located in the centre of the board where bending is at its maximum. The farther a device is from the centre, the more it can tolerate a higher number of falls; for instance, the devices placed in U6, U10 and U1, U5, U11, U15 are subject to a peeling stress which is five time smaller than the one acting on the central locations.

8.3.6 Effect of Different Parameters on Drop Test Results

Previous publications have investigated the effects of the variation of several parameters on the results of the test in order to better understand both the physics and the dynamics of the failures characterizing the impact. This section will describe some of the reported conclusions.

8.3.6.1 Fall Orientation

It has been experimentally proven that the worst case orientation for a falling board is the one where components are facing downwards, because in this case the joints are subject not only to the stress derived from the board bend, but also to the force of inertia generated after the impact.

8.3.6.2 Number of Screws Securing the Board

Both the location and the number of screws used to assemble the board inside the final product is an important factor, because it has a strong influence on the way the board vibrates, on the distribution of the deformations, and, ultimately, on the amount of stress acting on the joints. Figure 8.30a,b shows a comparison of the dynamic deformations a board is subject to during a drop test, both along the longitudinal and transversal directions, respectively, when either 4 or 6 screws are used (the two additional screws being located in the middle of the longer sides of the board).

It is clear that the vibration patterns and corresponding deformations depend on the number of mounting screws; for instance, the use of 6 screws results in a dominant deformation along the transversal direction. Using 4 screws results in a shorter time to failure, caused by the greater deformation induced on the board. Using a higher number of screws, board stiffness is increased, vibrations are inhibited, and the stress is reduced. Also, it is important to point out that by increasing the number of screws it is no longer possible to have a clear correlation between the failure of the devices and their position on the board: cracks will propagate randomly, because the distributions of both the deformation and the stress get much more complex.

8.3.6.3 Type of Material Used for the Joints

It has been proven that the eutectic alloy SnAgCu (one of the new lead-free alloys used in microelectronics) is characterized by a time to failure which is considerably shorter than that of the eutectic alloy SnPb used previously (i.e., before the introduction of the new European regulation which bans the use of lead because of its

Fig. 8.30 Effect of the number of screws on the amount of deformation along (**a**) longitudinal and (**b**) transversal direction

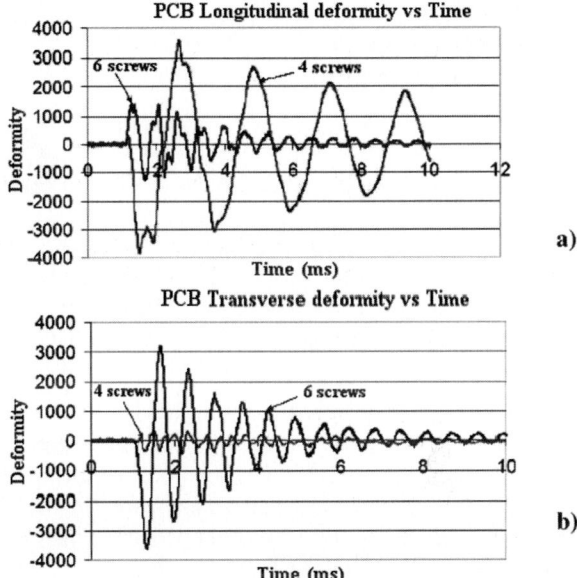

environmental impact and toxicity). SnAgCu alloys have a higher elasticity module; therefore, at a given value of deformation, and because of the direct relationship between stress and the elastic module, the peeling stress acting on the joints will be greater. The conclusions reached by analyzing which materials are most suitable to create joints for the drop test are exactly opposite the conclusions of the results of the thermal kind of tests; in this case SnAgCu alloys have proven to have a time to failure which is twice as much as the lead-based eutectic alloys, with regard to the same tested device. The fact that a device shows different performances when a drop test is applied compared to the case of thermal cycles is not completely wrong: the failure mechanisms for the two cases are totally different. The cracking of the joints during thermal tests is mainly caused by a cutting stress, and in general it occurs at the joint/copper-pad interface on the component side. In case of drop tests, the failure of the joints is mainly due to traction stress induced by the combined action of both mechanical shock and board vibration, and it causes a crack in correspondence with the joint/copper-pad interface which can occur either on the component or on the board side.

8.3.6.4 Size and Form Factor of the Board

In previous sections it has been proven that the outermost joints placed along the longitudinal direction of the board are subject to greater stress and therefore they tend to fail before the others. If the board is rotated by 90 degrees, the situation does not change: the joints placed along the new longitudinal direction of the board are more critical. Cracking of the joints is mainly caused by the vibration of the board.

Fig. 8.31 Distribution of deformations in a square board (48 × 48 mm) during drop test (bending upwards)

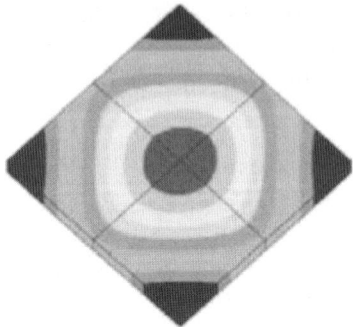

As a consequence, when using bigger boards, either rectangular or square, the stress acting on the joints induced by bending is greater, because the so-called distance to neutral point (DNP, i.e., the point where bending is zero) increases. Furthermore, it is a known fact that, in this case, the maximum bending on the centre line is proportional to the cube of the length of the board; therefore, the longer the board, the greater the deformations, and therefore stress is higher and time to failure is shorter.

Furthermore, it can be said that small, square boards should be preferred, in the case of drop tests, with respect to rectangular ones, because the stress is evenly distributed along the longitudinal and transversal directions, thus resulting in less severe stress for the joints (see Fig. 8.31). In this case too, greater deformations occur in the centre of the board and decrease moving towards the corners.

8.3.6.5 Characteristics of the Material of the Board

As far as the drop test is concerned, it is preferable to create the board using materials whose elasticity module has a higher value. This is due to the fact that a higher elasticity module implies a higher stiffness and therefore smaller bending stress.

8.3.6.6 Form Factor and Size of the Electronic Devices

Comparing devices having the same area, it can be seen that square components should be preferred over rectangular ones. As already seen for the boards, stress can be evenly distributed along longitudinal and transversal directions; while in case of rectangular components, the greater stress occurs along the longitudinal direction and concentrates on the corner joints, causing a shorter time to failure. The concept of DNP can be introduced in this case as well: inside a single device, the neutral point coincides with the barycentre or centre of gravity; therefore, the bigger the size (especially longitudinally), the greater the stress due to increased DNP.

This is another proof of the fact that rectangular devices have a worse performance. Furthermore, using square components it is possible to have a higher number of available joints, which allows a redistribution of the stress.

Lastly, in the case of drop tests, thinner components should be preferred, because they have a greater capacity of bending, and therefore the resulting stress will be less severe.

8.3.6.7 Characteristics of the Material Used for the Components

Components whose package is developed using low elasticity module materials show better performance in cases of drop tests, because a lower elasticity module implies a greater bending capability and therefore the material can better withstand the stress acting on it.

8.3.6.8 Effect of the Size of the Joints

Another important factor which influences the reliability of the joints is their size, and above all the so-called standoff of the device. Standoff is just the measure of the distance between the surface of the copper pad of the board and the package of the device mounted on it, evaluated after the soldering process. Standoff does not usually equate to the diameter of the joint, but it is in close relationship with it.

Both past experimental studies and the diagrams showing cumulative breaks can be used to infer that the greater the size of joint, the longer is the time to failure; by increasing the diameter of the joint from 0.3 mm to 0.4 mm, there is a 4.5x improvement.

8.3.7 Conclusions

In this section we have briefly presented the JEDEC international standard which describes the procedure to be used to execute a drop test, i.e., the test adopted to evaluate the reliability of solder joints used to mount a specific type of electronic device (in general, BGA-type memory devices) on integrated circuits (PCBs or boards). This test has also been described from both the mechanical and dynamic point of view; a brief analysis of both failures and related mechanisms leading to the cracking of the joints has been offered; and a description of the effects of the variation of some parameters on the test results has been reported.

During a drop test, which is mainly performed on the components to be used in portable electronic products, the cracking of the joints is caused by a combined action of mechanical shock and bending of the board; components mounted around the centre of the board are subject to greater stress compared to the outermost ones.

Inside a device, the critical joints are the ones located in the outermost corners, and they tend to crack at the interface (either on the component side or on the board side) between the copper pad and the joint in correspondence with the intermetal alloys which are produced during the soldering process. Both the start and the propagation of the cracking are caused by traction stress induced by the different bending stiffnesses which characterize the materials composing both the board and the package of the device. By comparing the two most common alloys used to create the joints, i.e., the classic SnPb alloy and the recently introduced SnAgCu, it has been proved that, because of the higher value of the elasticity module, the SnAgCu alloy has a worse behaviour during a drop test.

Last but not least, the failure mechanisms for drop tests and thermal tests are different; therefore it is important to identify which is the most useful test for the final application the device is used in. In the case of devices to be used in portable products, drop test performances are more critical than the ones for thermal tests.

For instance, in the case of mobile phones, the package must have a good resistance to impact, since the product may accidentally fall.

8.4 Thermal Cycling

The aim of the evaluation of the thermal cycle test on the board is to verify that the BGA and its corresponding soldering can tolerate a quick and significant thermal change; such an event triggers a mechanical stress caused by the difference between the coefficients of thermal expansion of either joint/component or joint/printed circuit. Thermal types of tests are of particular importance both for devices which are subject to significant changes in their ambient thermal condition across their lifespan, and for devices characterized by low power dissipation.

Even if mechanical types of tests (like bend and drop tests) have become more and more important lately, the recent introduction of the new lead-free alloys has increased the importance of the thermal types of tests, which have become almost indispensable to both understand and deeply investigate the performance of the new materials used in microelectronics. In fact, data are still scarce and limited concerning both the resistance to thermal stress of the lead-free joints and the impact of the kind of package used on the reliability of the joints.

8.4.1 Experimental Procedure

The test is usually performed inside an air-air environment in two different climatic rooms set at different temperatures. Two different conditions can be used, as shown in Table 8.7. In order to carry out the test, a proper board is used, onto which a daisy chain type of device is soldered.

During the test, the electrical resistance of the device is continuously measured, in order to easily detect any potential failure. In general, a failure is defined as an increase in the joint resistance up to 150–250 ohms.

Obviously such a value must be defined with respect to the device under test, that is with respect to the initial value of electrical resistance.

The test is considered successful if no failure occurs after 1000 thermal cycles. It is advisable to reach the number of cycles which cause failure for almost all the samples, in order to have enough data to perform a statistical analysis; such an analysis will allow a comparison of the performance of different devices and an evaluation of

Table 8.7 Conditions used to carry out thermal tests

Condition	Temperature Range (°C)	Temperature Range (°C)
	0 ~ 100	−40 ~ 125
Minimum Temperature	0	−40
Maximum Temperature	100	125
Temperature Variation	10 °C/min	11 °C/min
Time for Room	10 min	15 min
Total Test Duration	40 min/cycle	60 min/cycle

the effect of varying some parameters, such as, for example, the composition of the alloy used for the joints. The so-called *Weibull statistical distribution* is generally used to both describe and analyze the statistical distribution of the data collected during the test (the so-called *time to failure*). By using such a distribution, two main parameters can be derived, namely the *shape* and *scale* parameters, which can be used to compare different devices.

8.4.2 Analysis of the Failures

In general, the thermomechanical stress due to the difference in the thermal expansion coefficients between package and board is a cutting stress which will cause cracks in the most stressed point: that is, in the joints. Such failures at the joint level can sometimes be highlighted using just a simple optical microscope with either 10X or 50X magnification. Considering the way the joints are arranged in a device (which is usually defined as an *array*), the most critical zones, where cutting stress is at its maximum, are those zones in correspondence with the external perimeter. It must also be remembered that the soldering process of the device is of the utmost importance in a thermal type of test, also considering the failure mechanism; the presence of empty spots due to a bad degassing can accelerate the occurrence of the initial cracking and the following propagation of it. In fact it is well known that empty spots inside the joint must be considered as real defects, around which the stress concentrates; if the distribution of stress inside a joint is evaluated, it will turn out to be at maximum where the empty spots are. The high value of stress in such spots makes both the start and the propagation of the crack easier.

Other parameters which can have an influence on the thermal cycling tests are:

- *Effect of the size of the silicon die*: the larger the die, the worse the performance of the test. In general, the die decreases the average value of the coefficient of thermal expansion (CTE) of the device, which implies an increase in the difference between the CTE of the board and the CTE of the device, thus causing more stress.

- *Increase of the diameter of the joint*: by increasing the diameter of the joint, while keeping constant the solder mask opening, the time to failure or, better, the duration of the device, decreases. During the soldering process of the device on the board, the joints slightly collapse, and therefore the distance between the board and the device decreases; the overall structure will therefore suffer from a greater cutting stress.
- *Effect of the thickness of the substrate in the device*: increasing the thickness of the substrate improves performance. Using a proper substrate, placed inside the device, the total average value of the CTE of the package increases, thus reducing the difference between the CTE of the board and the CTE of the BGA.
- *Effect of the different temperature ranges*: the higher the difference between the maximum and minimum temperatures used to carry out the test, the greater the expansion of the materials. In fact, it has been determined that a temperature variation from 0 °C to 100 °C can cause more than a 20% improvement in reliability performance with respect to a temperature variation from −40 °C to 125 °C. It should be remembered that there are international standards which regulate the execution of these tests; the established rules must be followed in order to be able to compare the results fairly.

References

1. Darveaux R, Syed A Reliability of area array solder joints inbending. White Paper disponibile consultando il sito internet www.amkor.com
2. Electronic Industries Association of Japan (1992) Standard ofElectronic Industries Association of Japan: Mechanical stress test methods for semiconductor surface mounting devices
3. Shetty S, Lehtinen V, Dasgupta A, Halkola V, Reinikainen T (1999) Effect of bending on chip scale package interconnects. ASME International Mechanical Engineering Congress, Nashville
4. Leicht L, Skipor A (1998) Mechanical cycling fatigue of PBGA package interconnects. Proceedings of the International Symposium on Microelectronics
5. Juso H, Yamaji Y, Kimura T, Fujita K, Kada M (2000) Board levelreliability of CSP. Proc IEEE ECTC
6. Perera UD (1999) Evaluation of reliability of µBGA solder joints through twisting and bending. Microelectron Reliab No. 39
7. Rooney DT, Castello NT, Cibulsky M, Abbott D, Xie D (2004) Materials characterization of the effect of mechanical bending on area array package interconnects. Microelectron Reliab No. 4
8. Tee TY, Ng HS, Luan J, Yap D, Loh K, Pek E, Lim CT, Zhong Z (2004) Integrated modeling and testing of fine-pitch CSP under board level drop test, bend test and thermal cycling test. ICEP Conference, Japan
9. JEDEC (2003) Board level drop test method of components for handheld electronic products. Standard JESD22-B111
10. JEDEC (2001) Mechanical shock. Standard JESD22-B104-B
11. Chiu A, Pek E, Lim CT, Luan J, Tee TY Zhong Z (2004) Modal analysis and spectrum analysis of PCB under drop impact. SEMICON Singapore
12. Pek E, Lim CT, Tee TY, Luan J, Zhong Z (2004) Novel numerical and experimental analysis of dynamic responses under board level drop test. EuroSIME Conference, Belgium

13. Pek E, Lim CT, Tee TY, Luan J, Zhong Z (2004) Advanced experimental and simulation techniques for analysis of dynamic responses during drop impact. 54th ECTC Conference, USA

14. Abe M, Kumai T, Higashiguchi Y, Tsubone K, Mishiro K, Ishikawa S (2002) Effect of the drop impact on BGA/CSP package reliability. Microelectron Reliab No. 42

15. Ng HS, Tee TY, Zhong Z (2003) Design for enhanced solder joint reliability of integrated passives device under board level drop test and thermal cycling test. 5th EPTC Conference, Singapore

16. Luan J, Yap D, Loh K, Pek E, Lim CT, Tee TY, Ng HS, Zhong Z (2004) Integrated modeling and testing of fine-pitch CSP under board level drop test, bend test, and thermal cycling test. ICEP Conference, Japan

17. Lim CT, Pek E, Tee TY, Ng HS, Zhong Z (2003) Board level drop test and simulation of TFBGA packages for telecommunication applications. 53rd ECTC Conference, USA

18. Huang CY, Liao CC, Zheng PJ, Hung SC, Wu JD, Ho SH (2002) Board level reliability of a stacked CSP subjected to cyclic bending. Microelectron Reliab No. 42

19. Tee TY, Ng HS (2000) Board level solder joint reliability modeling on TfBGA package. International Conference on Electronic Reliability

Chapter 9
Reliability in Wireless Systems

A. Chimenton and P. Olivo

9.1 Introduction

Cell phones represent a quite recent personal communication system, dating back only about 25 years. Initially the installation costs of the network infrastructures, the service, and the cellular phones, limited their use to commercial and professional customers and their service coverage to big cities in countries with major industrial development. A fast and growing technological development, together with an upsurge in customer demand, allowed both an increase in coverage and a reduction in service costs. These effects have further widened the total market opportunity for wireless devices. The introduction in 1993 of digital communications and of the GSM standard led off an overwhelming development of mobile communications; in addition to the first common commercial usage, a new consumer market started showing up; besides the principal voice services, new functions have begun to appear, such as the short messages service (SMS), introduced to solve providers' service needs and now one of the principal income sectors for the mobile communication providers.

The recent introduction of new standards (GPRS and UMTS) increased considerably the applications related to cell communications, such as transmission of data, images, music, and video, thus making voice transmission just one of many system applications, and not necessarily the most remarkable.

Cell phones, or more properly, "mobile terminals," followed, and often even drove, the development of digital wireless communications. At first the manufacturers' goals included weight and dimension reduction, the increase of the voice signal to noise ratio, and the reduction of power consumption in order to increase battery duration. Afterwards, the mobile terminals market has been driven by different factors, such as screen dimensions and the possibility of making pictures or videos. These simple considerations, well known even to normal users following the

Andrea Chimenton
University of Ferrara, Ferrara, Italy

Piero Olivo
University of Ferrara, Ferrara, Italy

R. Micheloni et al. (eds.), *Memories in Wireless Systems*,
© Springer-Verlag Berlin Heidelberg 2008

evolution of digital wireless communications, are strongly related to the subject of the present chapter: the reliability of mobile terminals for wireless communications.

Next to the introduction of digital wireless communications, it is difficult to find an object with the same status symbol value as a mobile terminal. In many social environments it is mandatory to own and show the latest model with the most innovative multimedia functions, and then replace it with a new one, with additional functions, just three or four months later. As a consequence, cell manufacturers are continuously forced to introduce new models and new functions so that, in that context, *"time to market"* importance becomes absolutely incomparable to that of other systems in the electronics consumer market. The time to market reduction obviously impacts the new system design time, as do prototyping, testing, and engineering. Of course, the fast time to market may not disregard the reasonable confidence that the product will operate correctly, since any problem in that area may result not only in the flop of the product, but also, more generally, in distrust and loss of corporate image, with potentially negative effects on the overall range of products. Furthermore, it is not sufficient to guarantee that performance and characteristics are present when the product is sold; rather, it is necessary that they remain operative for a sufficient time (the operative life). Reliability, defined in a quantitative way as the probability that an object will fulfil a determined function in certain conditions and for a given period of time, is, at least for consumers not driven by image considerations, a product evaluation parameter at the same level of importance as its functional characteristics.

In the area of cell phones, the verification of functions in new models and the estimation of long term reliability assume a different meaning compared to other electronic equipment; this is particularly true regarding the operating lifetime. It is important to observe that the lifetime of a mobile terminal hardly coincides with the time in which it operates correctly with no critical failures. Often, the time to be considered is equivalent to the product's commercial life (the time that elapses before a still operating product is replaced by a more innovative one), or at least to the battery lifetime (in many cases battery replacement may not be as convenient as replacement of the entire cell phone).

In the cell market a basic parameter to be considered is the average time occurring before the first failure (*mean time to failure*, MTTF), while other parameters characterizing a system's reliability, such as the *mean time between failures* (MTBF), have no practical use. In particular, it is fundamental for the customer that the first failure does not occur before the product's anticipated replacement with a new, more innovative device. In the case of failure, since cell phones are considered essential goods, the customer will acquire a new model, and his choice will be driven by considerations of mistrust against the brand of the failed mobile phone.

For a manufacturer, it is mandatory that the mobile apparatus will operate for the entire warranty period, which, as has emerged from the previous considerations, for a large set of customers (in particular those more focused on market novelties), extends over the commercial lifetime. For the manufacturer repair costs could be too high, particularly for fault detection; if a failure cannot be immediately located, it is more convenient to replace the entire phone board rather than to identify and locate the failure with the aim of replacing only the failed block. The reliability of an electronic system must be continuously verified during the development and production cycle by means of suitable testing phases. Tests verify the product's operation for a given period of time under given operating conditions. In such a way it is possible to estimate some parameters representing the reliability level, such as the MTTF.

A main problem for the testing phase is its duration. Time to market can benefit from production cycle shortening; whereas a longer time spent in reliability improvements may give rise to a higher MTTF for the product. The estimation of the reliability parameters becomes more and more difficult when joined to shorter development and production time. Specific methodologies based on accelerated tests are then used to execute reliability control tests in relatively short times. The use of extreme operating conditions (high temperature/humidity, vibrations, high operating voltages/currents) allows extrapolating the reliability characteristic parameters compared to normal operating conditions. It must be observed that a wireless terminal, by its nature, must operate under different environmental conditions (outdoors, on a vehicle, etc.), and is therefore prone to the presence of humidity, large temperature fluctuations, vibrations, and shocks, much more than any other commercial product.

Accelerated tests, besides assisting the achievement of product specifications, also allow foreseeing reliability problems that therefore can be tackled in the production phase. This concept is fundamental, since the maintenance/modification cost increases as a function of product maturity (see Fig. 9.1). For these reasons, reliability issues are today taken into account at earlier design phases, in the so-called *design for reliability* methodology. The product design flux therefore assumes the aspect depicted in Fig. 9.2, where it can be observed that reliability is included among the design parameters together with cost and performance.

By taking reliability into account from the first stages of the design phase, the overall design cost increases. It is therefore essential to find the best trade-off between design and maintenance costs, so that a minimum total cost can be realized, as shown in Fig. 9.3.

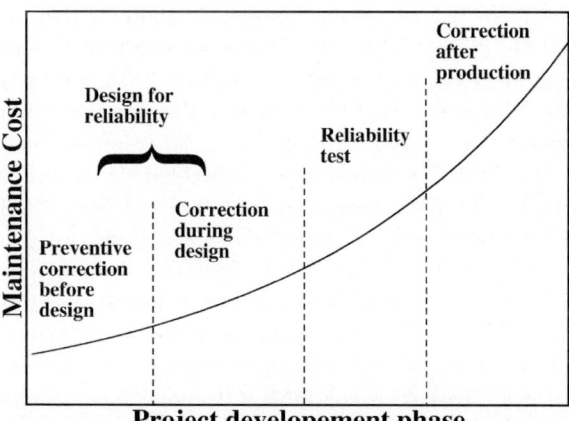

Fig. 9.1 Dependence of the maintenance cost on the project development phase

Fig. 9.2 Top-down project
flux taking into account
reliability issues

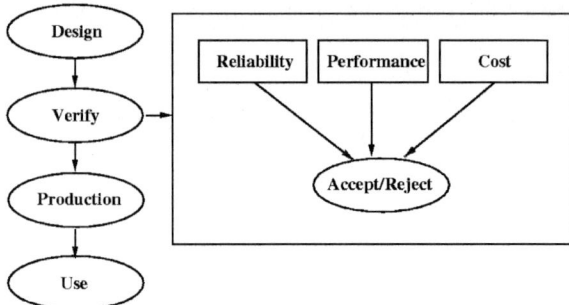

Fig. 9.3 Trade-off between
maintenance costs and project
costs connected to reliability

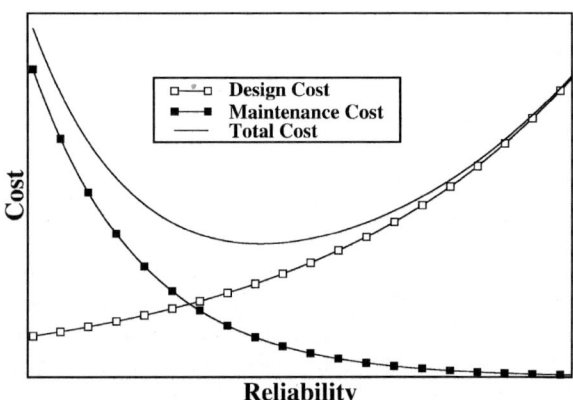

For wireless systems, too, the use of methodologies aimed at achieving predetermined reliability targets is spread along the entire development cycle, from the idea to the mass production.

Table 9.1 shows the description of the different phases, the objectives related to reliability, and the instruments used to achieve them.

Figure 9.4 shows the product reliability improvement (*mean time to failure*, MTTF) during the different development phases. It can be observed that no significant improvements are present at phase 1, since the project is still at an ideal stage. Phase 2, on the contrary, evidences a large reliability improvement. Improvements begin to slow down during phase 3, and eventually end in phase 4.

The objective of this chapter is twofold: on the one hand, techniques and methodologies of failure science are introduced for each development phase; on the other hand, some practical examples of these methodologies are shown for the case of wireless systems. Figure 9.5 summarizes a cellular system block diagram. It is composed of several different subsystems (battery, antenna, semiconductor memories, etc.). The figure also shows the blocks whose reliability aspects will be described in the following sections as practical examples of the general methodologies of failure science. The last row of Table 9.1 reports the examples considered for each development phase.

Table 9.1 Description of development phases and methodologies implemented to improve reliability

	Phase 1 idea	Phase 2 validation	Phase 3 development	Phase 4 production
Description	Operative specs determination	Materials and components choice	Prototypes development	medium/large scale production
Reliability objectives	Individuation of failure mechanisms and modes	Reliability improvement (at least 60%)	Reliability demonstration and validation with a certain confidence level (90%)	Screening of latent defects, monitoring
Reliability instruments	FMEA RPM Benchmarking	TAAF, HALT Strife	Test at zero faults, Accelerated tests	Burn–in
Considered examples	MESFET, Flash memories, corrosion, ionic migration	Vibrating motor	Cellular, vibrating motor	row/column redundancy in memories

During phase 1 the reliability of the entire system can be estimated in a general way, thanks to failure analysis predictive tools based on failure probability data/ models available for each system component (*reliability predictive modeling*, RPM). The resulting model can be semi-empirical and, as such, will be based on a huge amount of data. The Military Handbook is a standard example of this type of predictive methodology. Alternatively (or in addiction), it is possible to take into consideration the physical knowledge of failure mechanisms that are always supposed to be present. The resulting models allow calculating the MTTF in a more accurate way as a function of some basic physical quantities involved in the failure mechanisms.

The knowledge of both kinds of models is fundamental to address the first project phases and provides accurate estimates of system reliability. In addition, the failure models allow selecting the methodologies, the criteria, and the characteristic parameters for the accelerated tests performed during the design and validation phases.

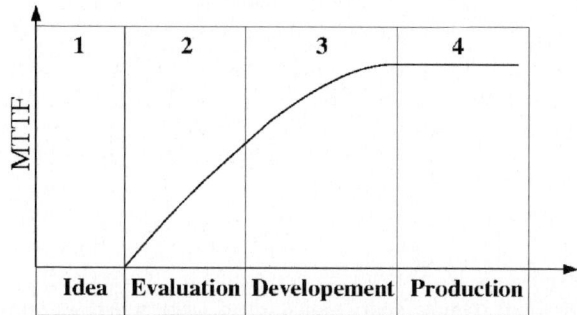

Fig. 9.4 Reliability improvement at different development phases

Fig. 9.5 Block diagram of a
typical mobile phone

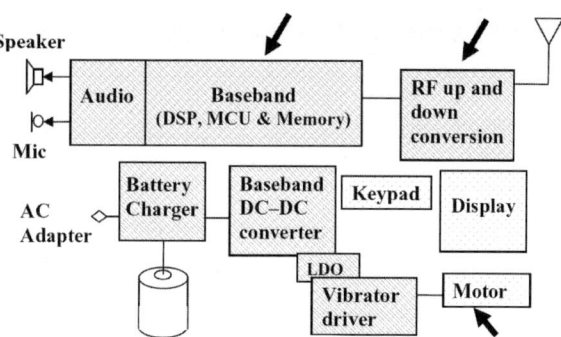

Examples of instruments used during phase 1 and related to reliability prediction
and modelling will be discussed in Sect. 9.2; they will include electronic devices
such as MESFET and semiconductor memories, and some physical mechanisms
such as corrosion and ionic migration. The evaluation of the MTTF will also be
addressed.

Phases 2 and 3 will be discussed in a more general way in Sects. 9.3 and 9.4,
where the techniques used in these phases will be applied to a common mobile
phone part: the vibrating motor. Issues related to phase 4 will be tackled in Sect. 9.5
regarding the burn-in, the fault tolerance, the relationships between defects, yield,
and reliability, and the use of redundancy in memories.

9.2 Reliability Prediction Models (Phase 1)

9.2.1 Memory Reliability in Wireless Systems

The reliability study at an early stage of a project is based on knowledge of the
reliability of each single component. If a system is constituted by n blocks, each
of them correctly operating to keep the system alive, then the system reliability
$R_{system}(t)$ (the probability that the system operates correctly in the time interval
from 0 to t) is given by:

$$R_{system}(t) = \prod_{i=1}^{n} R_i(t), \qquad (9.1)$$

where $R_i(t)$ is the single block reliability.

There are several prediction models which allow estimating the component
failure probability in an electronic system as a function of some design parameters.
Among them, the MIL-HDBK-217 is the most common. Such a standard, developed
by the U.S. Defense Department, is based on a huge quantity of failure data collected
by specific measurements performed on specific technologies. These measurements
can last several years, causing an issue when systems include relatively new or

updated components. The MIL-HDBK-217 models rely on a basic hypothesis: the failure rate of a single component of a system is assumed to be constant in the course of time. Two major implications derive from such an assumption: on the one hand, the reliability calculation of a system is simplified, but on the other hand, the analysis is limited to the device's expected life. The first implication is evident from the following considerations. It can be demonstrated that the hypothesis of a constant failure rate for all components is equivalent to considering a reliability function which varies exponentially with time, that is:

$$R_i(t) = e^{-\lambda_i t},$$ (9.2)

where λ_i is the failure rate of the i-component.

The reliability of the whole system is thus given by

$$R_{system}(t) = e^{-\sum_{i=1}^{n} \lambda_i t} = e^{-\lambda_{tot} t},$$ (9.3)

where λ_{tot} is the failure rate of the entire system.

It follows also that the entire system is characterized by a constant failure rate that is given simply by the sum of the failure rates of the single components. Therefore, for such a system, MTTF is given simply by:

$$\frac{1}{MTTF_{system}} = \lambda_{tot} = \sum_{i=1}^{n} \lambda_i = \sum_{i=1}^{n} \frac{1}{MTTF_i},$$ (9.4)

where $MTTF_i$ is the mean time to failure of the i-component.

The second implication refers to the so-called bath tube curve. Typically, a product failure rate tends to follow a trend similar to that shown in Fig. 9.6, where three phases can be observed: infant mortality, useful life, and wearout. In the infant mortality phase the failure rate decreases until the defective population runs out completely. In the useful life phase the failure rate is characterized by a constant

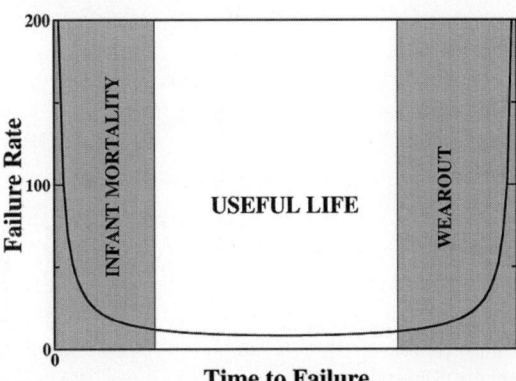

Fig. 9.6 Failure rate trend along the entire product life

value. In the wearout phase the failure rate increases with time, since product aging increases failure probability. The manufacturer is interested in selling products that do not exhibit early failures. Screening techniques are able to detect the defective population, so that products that are beyond the infant mortality phase can be sold. The warranty period must fall in the useful life period, which is characterized by a constant failure rate, and must not enter the wearout area. The results derived from the Military Handbook standard, based on constant failure rates, are therefore referred only to the useful life period of a product.

The reliability prediction methodology based on the MIL-HDBK-217 will be described by means of a practical example concerning the failure rate of semiconductor memories used in mobile applications and their dependence on specific factors, such as power consumption and number of write cycles. Flash memories, widely used in wireless systems, are characterized by high performance and scalability combined with relatively low costs. Figure 9.7 shows the trend of the significant penetration of Flash memories in the cellular phone market. The rapidly increasing complexity of Flash memories, however, makes it more critical to guarantee a predetermined level of reliability. Table 9.2 shows the failure rate of some components present in electronic systems; memories belong to the integrated circuit (IC) category and present relatively high failure rates because of their intrinsic complexity.

Following the MIL-HDBK-217 model, the failure rate of a memory for mobile applications, expressed as number of failures every 10^6 hours, is given by:

$$\lambda_p = \left(C_1 \pi_T + C_2 \pi_E + \lambda_{cyc}\right) \pi_Q \pi_L, \tag{9.5}$$

where C_1 is a parameter depending on (following an expression reported in the MIL-HDBK-217): memory capacity, worst case junction temperature, activation energy, package temperature, power consumption, and, finally, package thermal resistance. Package failures are included in C_2, which depends on the number of pins.

Other factors to be specified in the failure rate expression are:

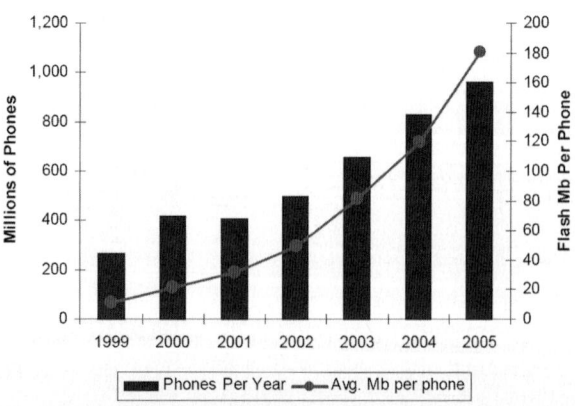

Fig. 9.7 Flash memories in cellular phones (source: Semico Research Group)

Table 9.2 Failure rate in FIT (Failure in Time over 10^9 hours—Components) for some electronic components

Capacitors in mylar	Electrolytic Ta solid	Electrolytic Al	Connectors (per pin)	Resistors	Transistors Si npn, 1W	Welding	IC, 30 gate	IC, 100 gate
1–3	3–8	10–40	0.5–1.5	0.5–2	8–20	0.1–0.3	20–60	30–100

- π_E, environment factor that, for wireless systems, belongs to the ground mobile (GM) type;
- π_L, learning factor, that can be considered equal to 1 for mature technologies;
- π_Q, technology quality factor.

Finally, λ_{cyc} is a factor proportional to the number of programming/erasing cycles. It also depends on the availability of error correcting codes.

Technology scaling and the integration of a rising number of modules on the same die to meet the new requirements imposed by multimedia applications (pictures, videos, MP3, etc.) induce an unavoidable increase in power consumption, memory capacity, number of pins, and number of writing cycles. It can easily be understood that the model considers an increasing failure rate for all these cases, and in particular for the number of writing cycles, capacity, and power consumption (see Fig. 9.8).

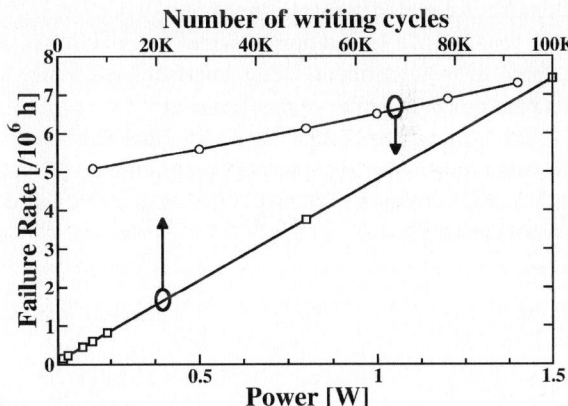

Fig. 9.8 Failure rate as a function of the number of both writing cycles and power consumption

9.2.2 MTTF Physical Models of Components and Interconnections

9.2.2.1 MESFET

The following study of failure physics models will consider, as an example, the MESFET device that is the basis of several other devices (HEMT, PHEMT, etc.) which are frequently used in radiofrequency circuits of communication systems. Its relative simplicity allows a simple introduction to failure physics. The MESFET

Fig. 9.9 MESFET cross
section

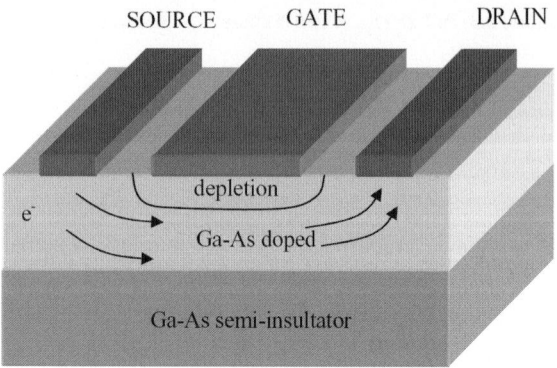

(see Fig. 9.9) is essentially constituted by two ohmic contacts (drain and source) and by a Schottky contact (gate) fabricated over a doped substrate of a compound semiconductor (gallium arsenide, GaAs).

The MESFET operation relies on the control of the channel conductivity by means of a gate voltage. In practice, the gate voltage narrows/enlarges the depleted region below the gate, thus reducing/increasing the conductivity between source and drain. This effect allows, in practice, electrical control of the channel conductivity.

Failures may afflict several MESFET regions: the metallic gate (corrosion), the drain/source ohmic contacts, the Schottky barrier, and the free surface between gate and source/drain (migration, channel hot electrons); see Fig. 9.10. The problems related to metal/semiconductor interfaces are strongly dependent on temperature and are based on diffusive mechanisms.

Short circuits or broken lines and abnormal behaviors may stem from electron/ion migration in the free space between source/drain and gate metallization and from gate metal corrosion. The insulation oxide can be affected by trapping phenomena of hot carriers.

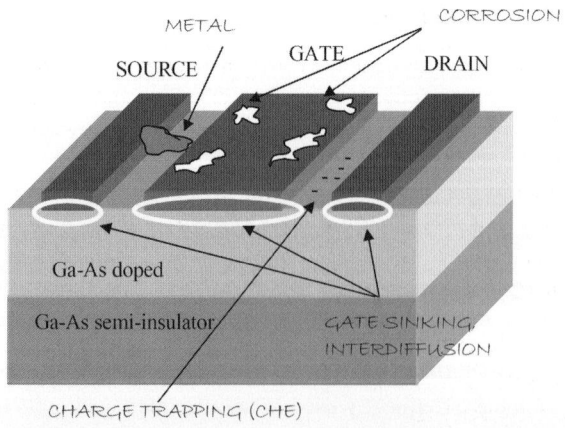

Fig. 9.10 MESFET regions
involved in failure
mechanisms

9.2.2.2 Gate Region and Schottky Barrier

The Schottky barrier can be affected by the "gate sinking" phenomenon, that consists in the gate metal diffusion into the GaAs substrate. Among the principal effects of gate sinking, the following can be recalled: drain current reduction at zero gate bias voltage because of the reduction of the active channel thickness, pinch-off voltage reduction, and channel degradation. It is possible analytically to derive a degradation model for the pinch-off voltage (V_p) caused by gate sinking, starting from simple considerations of metal diffusion in a semiconductor. The result gives the following relationship:

$$\Delta V_p(t, T) = 5.5 \frac{qNW}{\varepsilon} \sqrt{D_0 t} e^{-\frac{E_d}{2kT}}, \tag{9.6}$$

where q is the elementary charge, N the semiconductor doping, W the channel width, ε the dielectric constant, D_0 the metal diffusion constant into GaAs, E_d the diffusion activation energy, t the time, T the temperature, and k the Boltzmann constant. The degradation is thus a parabolic function of time, and it is thermally activated. Typical activation energies in Al contacts are in the 0.8–1.0 eV range. Improvements in gate metallization technologies in the last two decades have allowed an extension of the component life from a few thousands of hours up to millions of hours by moving from an Al technology to titanium, gold or platinum alloys. Among them, titanium is the most used material. It requires a gold covering to solve wire bonding problems at its surface, and an interdiffusion barrier to solve instability problems at temperatures higher than 200 °C. Therefore, the resulting structure is of Au/Ti/barrier/GaAs type. The barrier is made of Pt, W, Pt/Cr, or Pd alloys. The Au diffusion into GaAs through the metal barrier layers is one of the basic failures in gold metalizations. The activation energy of this mechanism, however, is about 1.5–1.8 eV. Therefore, Au/Ti/barrier/GaAs contacts present a greater stability compared to the simpler Al/GaAs contacts.

9.2.2.3 Metal Migration

Metal migration and its consequence, i.e., interconnection degradation, are extremely important phenomena in portable circuits exposed to high humidity and pollution levels. Metal migration consists in an electrochemical phenomenon occurring at room temperature in the presence of low leakage currents. Metals migrate from one interconnection towards another because of the presence of an electric field and a layer of water solution, which creates characteristic ramified structures (dendrites) constituted by the migrated metal. These structures can provoke a short circuit between the two lines. The most frequent failure mechanism, however, is caused by an open circuit created in the line from which the metal migrates.

A simplified model for this phenomenon gives the following relationship for the mean time to failure (presence of an open circuit in a line) as a function of the conductivity and of the amount of the water solution:

Fig. 9.11 Ionic migration between two metal layers

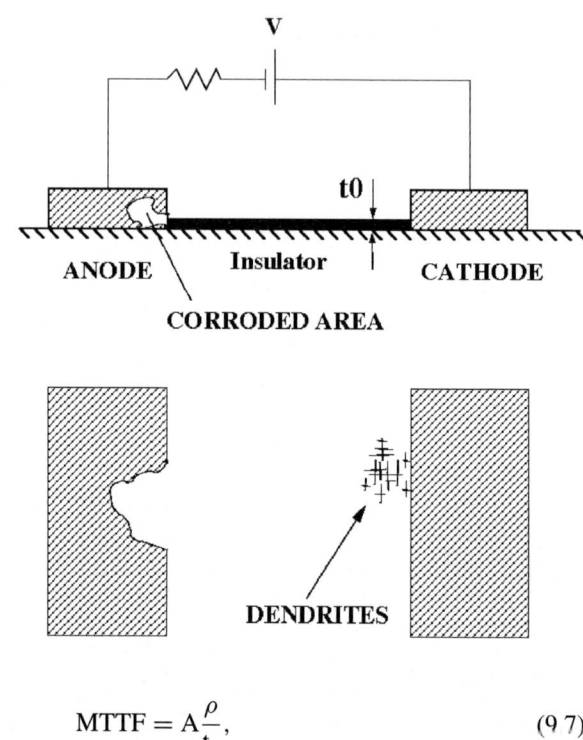

$$\text{MTTF} = A\frac{\rho}{t_0},\tag{9.7}$$

where A is a technology and metal dependent constant, ρ is the solution conductivity, and t_0 is the solution thickness (see Fig. 9.11).

9.2.2.4 Interconnection Corrosion

Corrosion may be triggered by the percolation of humidity, chlorine, or other chemical elements present in polluted environments. The percolation through the passivating layer protecting integrated circuits may be the consequence of the formation of small cracks or fractures caused by huge vibrations, collisions, thermal cycles, or latent defects. For a given relative humidity RH close to the interconnection (the effective RH value is time-dependent following the percolation kinetics), the mean time to failure because of the corrosion of aluminum interconnections is given by:

$$\text{MTTF} = \frac{A}{V^m(RH)^n}e^{\frac{E_d}{kT}},\tag{9.8}$$

where A, m, and n are constant, E_d is an activation energy for the chemical reaction, and V is the voltage applied to the line. This equation is frequently used in the design of interconnections.

9.3 Reliability Improvement (Phase 2)

The reliability of a product is the probability that it will maintain its functional characteristics for a given period of time. MTTF is one of the parameters that reliability science tries to calculate and improve. The most significant improvements can be reached during the product evaluation phase, thanks to *test analyze and fix* (TAAF) methodologies: the observed failures are physically inspected (failure analysis) and specific correcting actions are introduced to eliminate recurrent errors and to guarantee product improvement before manufacturing. The TAAF methodology is based on failure data provided by several accelerated tests designed to identify all possible reliability problems by stressing the product under extremely critical operating conditions, beyond the limits imposed by the design specifications. In the following, we will describe two types of tests: HALT and Strife.

9.3.1 HALT (Highly Accelerated Life Testing)

HALT is a well consolidated methodology that uses a chamber where samples undergo different types of tests and/or combinations of tests (basically temperature and vibrations). These tests are performed during the initial developing phase. The "four corners" HALT is frequently used: two tests are performed, one for temperature and one for vibrations. For each test the operation/breakdown boundaries are detected and therefore the four corners of the breakdown area are determined.

Units under test can still operate with temperature and vibration levels belonging to the gray area in Fig. 9.12, i.e., the area between the operating region and the breakdown boundary, although not all specs are necessarily verified. Also, from this region it is always possible to switch back to the operating region in a reversible way. On the contrary, when the breakdown boundaries are reached, the characteristics of the unit under test are associated with a permanent degradation and with failures whose peculiarities are to be analyzed and studied. Temperature testing allows detecting failure modes related to mechanisms that do not depend on humidity, such as,

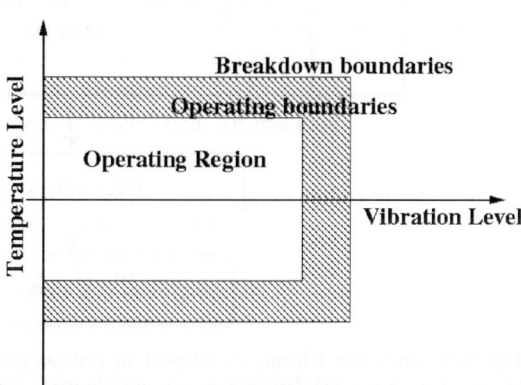

Fig. 9.12 The four corners for the operating/breakdown areas determined by the HALT test

condition $MTTF_{real} > MTTF_{target}$ is assumed as real, that is, the mean time to failure satisfies the target specification.

By considering a sample of n units, the probability that a test of duration t produces a maximum number c of failures is given by:

$$P(t) = \sum_{x=0}^{c} \binom{n}{x} (1 - F(t))^{n-x} F(t)^x. \tag{9.14}$$

In the case of a zero failures test ($c = 0$) and of a Weibull distribution, the probability of observing zero failures is simply given by:

$$P(t) = (1 - F(t))^n = e^{-n\left(\frac{t}{\alpha}\right)^\beta}. \tag{9.15}$$

There are two possible outcomes of the test:

1) if the test produces zero failures, $MTTF_{real} > MTTF_{target}$ is assumed;
2) if the test produces at least one failure, no assumptions are made.

The probability of a wrong assumption in case 1) is given by the probability of observing zero failures if $MTTF_{real} < MTTF_{target}$. This probability will be necessarily lower than the probability expressed in Eq. (9.15), and the manufacturer wants it to be as low as possible. The following condition is then obtained:

$$1 - L = e^{-n\left(\frac{t}{\alpha}\right)^\beta}, \tag{9.16}$$

where $1 - L$ is the so-called manufacturer risk.

By considering a level of confidence L (for example, 95%), it is possible to determine the test duration t as a function of the number n of samples:

$$t = \alpha \left(-\frac{\ln(1 - L)}{n}\right)^{\frac{1}{\beta}}. \tag{9.17}$$

Figure 9.15 shows the relationship between the test duration and the number of samples for different levels of confidence; for a given level of confidence, it is possible to reduce the test duration by increasing the number of samples. The test allows one, therefore, to establish whether the mean time to failure ($MTTF_{real}$) is higher than an expected mean time to failure ($MTTF_{target}$) with a given confidence level. By assuming, for example, an $MTTF_{target} = 80$ hours for normal operating conditions, then, for a test condition featuring $T = -20\,°C$ and $V = 5\,V$, the $MTTF_{target}$ is equal to 10.7 hours (see Eq. (9.13)).

The failure distribution during the test will follow a Weibull distribution with the following parameters: $\beta = 2.9$ and $\alpha = 10.7/\Gamma(1 + 1/2.9) = 12$.

Fig. 9.15 Test duration as a function of the number of samples for different confidence levels

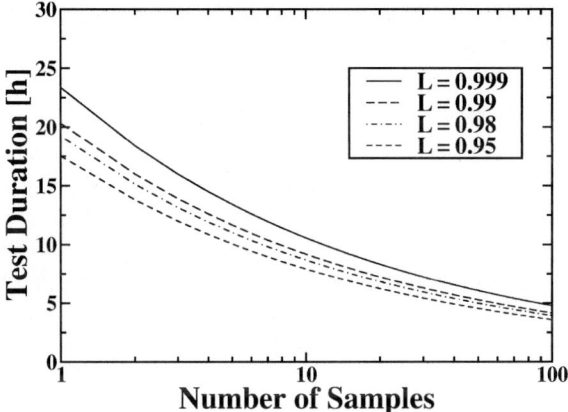

Using Eq. (9.17) with these values of α and β and assuming n $=$ 10, a test time of t $=$ 10 hours is obtained. Table 9.3 shows that the first failure occurred after 8 h $<$ 10 h, therefore the zero test failed and nothing can be said regarding $MTTF_{real}$ with respect to the target value of 80 hours. With a $MTTF_{target} = 60$ hours, $MTTF_{target}$ under the considered stress condition is 8 hours. The Weibull parameters are $\beta = 2.9$ and $\alpha = 9$ and the test duration at zero failures is 7.5 hours. Therefore, it is possible to be confident at the 95% level that $MTTF_{real}$ is at least larger than 60 hours (if it is lower, the probability of having a test at zero failure in t will be $<$ 5%). Finally, these equations allow establishing the optimal number of samples minimizing the cost C associated with the test activity (for example, see Fig. 9.16):

$$C = tc_1 + nc_2 = c_1\alpha\left(-\frac{\ln(1-L)}{n}\right)^{\frac{1}{\beta}} + nc_2, \qquad (9.18)$$

where c_1 and c_2 are constants.

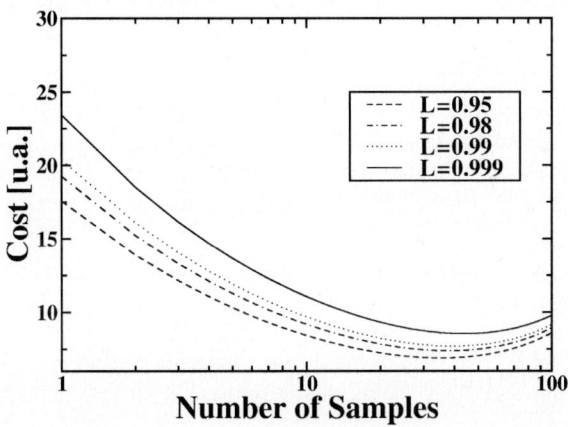

Fig. 9.16 Test cost as a function of the number of samples for different confidence levels

9.5 Screening/Burn-In, Defectiveness and Fault Tolerance (Phase 4)

Scaling of circuit dimensions allows integrating on a single chip the functionalities previously performed by separate modules. In this way it is possible to eliminate the external interconnections among different modules, that may be prone to failures related, for example, to the quality of solders, packages, and interconnections (corrosion, electromigration, etc.). The reduction of the line capacitive loads increases the system performance and, in general, the reliability at the package level. Nevertheless, because of the increasing number of functionalities that are implemented, the reduction of device dimensions is often associated with larger chip/modules also. Assuming constant defect densities and wafer dimensions, the larger the chip, the lower is the resulting yield (see Fig. 9.17).

It must be stressed that miniaturized devices are more sensitive to defects of smaller dimensions (the critical defect dimension decreases). Therefore, even with the same module area, the number of potentially dangerous defects is larger. This number, in addition, increases more than linearly as critical dimensions decrease, since the defect densities D vary with respect to their diameter x as

$$D = \frac{k}{x^3}. \tag{9.19}$$

Line cleanness, process control, and the use of redundancy techniques become more and more important to avoid yield reduction and cost increase, as well as reliability degradation. A basic correlation between yield and reliability exists: productive cycles with low defectiveness usually give rise to a higher reliability.

Figure 9.18 shows schematically how defects represent the link between yield and reliability. It must be observed that reliability and yield refer to completely different situations: reliability is concerned with failures occurring during the product operation; whereas yield is related to not-operating (defective) products, that can be detected at the end of the development cycle. Reliability depends on operative

Fig. 9.17 Yield reduction following an increment of the chip dimensions for the same defect density (44 defects distributed along the wafer): on the left there are 44 defective chips over 492 (yield = 91%); on the right there are 33 defective chips over 115 (yield = 71%)

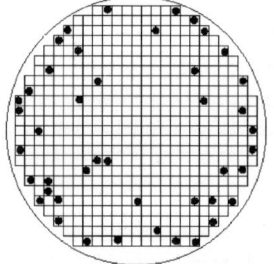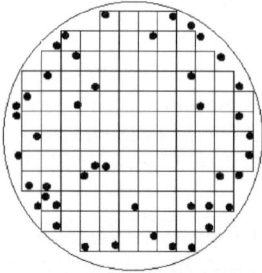

Fig. 9.18 Correlation between yield and reliability via defects

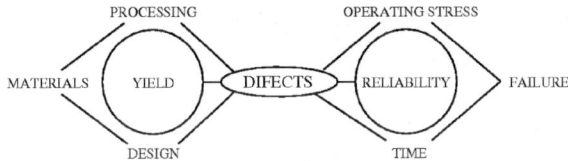

stresses and time. Yield, on the contrary, is a function of production process, materials, and design.

The strong correlation between yield and reliability means that it is extremely important to detect all defective components before selling. This statement holds in particular for the mobile phone market, since, as was already mentioned in the introduction, early failures play a dominant role because of short commercial life. Screening, burn-in, and fault tolerance are techniques normally used to detect defective populations, and they will be briefly analyzed in the next paragraphs.

9.5.1 Correlation Between Yield and Reliability

Circuit integration plays a fundamental role in the overall system dimension reduction and performance improvement of wireless systems. To reach this goal it is necessary to solve nontrivial problems related to yield. As a first approximation, the yield of a process producing modules of area A with a defect density D is given by:

$$Y = e^{-DA}. \tag{9.20}$$

Because of the continuous module area increment, it is necessary to reduce the defect density to keep acceptable yield levels. This goal can be achieved by improving the process quality and cleanness. The need for high levels of yield basically derives from economic considerations. It is known, however, that products featuring high levels of yield also reveal a high reliability, since defects reducing yield often have the same nature as those reducing reliability. Let's consider the example of Fig. 9.19.

In case 1 the defect creates an open circuit that can be detected during device testing and therefore produces a yield decrease. In case 2 the defect has smaller dimensions, it does not produce an open circuit that can be detected during testing, and it does not affect the process yield. The product is sold, but probably its lifetime will be relatively low. In fact, latent defects can give rise to an open circuit due to, for example, electromigration phenomena which are caused by an excessive current density. For a given distribution of defect densities, for example that of Eq. (9.19), it is possible to derive the mean failure probability P:

Fig. 9.19 Effect of a defect on an interconnection: 1) the defect produces an open circuit that will reduce the process yield; 2) the defect does not reduce the yield but it can affect the interconnection reliability

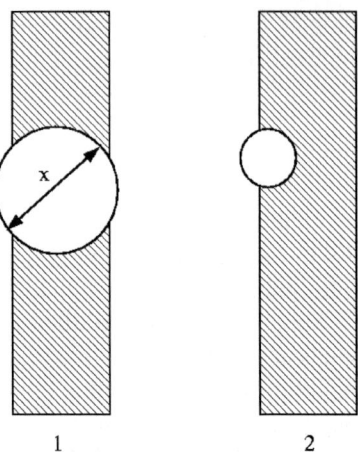

$$P = \frac{\int\limits_0^\infty P_Y(x)D(x)dx}{\int\limits_0^\infty D(x)dx}, \tag{9.21}$$

where P_Y is the probability that a defect with diameter x can induce a failure affecting the yield (open circuit). Supposing that:

$$P = \bar{D}A_Y, \tag{9.22}$$

where A_Y is the so-called critical area and:

$$\bar{D} = \int\limits_0^\infty D(x)dx, \tag{9.23}$$

the yield becomes:

$$Y = e^{-P} = e^{-\bar{D}A_Y}. \tag{9.24}$$

Similarly, the reliability R can be expressed in terms of the critical area A_R as

$$R = e^{-\bar{D}A_R}. \tag{9.25}$$

It follows that the fundamental relationship between yield and reliability can be expressed as:

$$R = Y^{\frac{A_R}{A_Y}}. \tag{9.26}$$

Equation (9.26) has been verified experimentally and has produced, for example, a critical ratio equal to:

$$\frac{A_R}{A_Y} = 0.3. \tag{9.27}$$

9.5.2 Fault Tolerance

Fault tolerance in VLSI circuits, particularly in wireless systems, may be obtained by exploiting redundancy techniques that duplicate specific system modules: subarrays in memories, computing elements in vector processors, macros in other digital circuits.

Redundant modules are used to replace defective modules detected during the screening phase. The substitution is done by reconfiguring the device. A classic example derives from the row redundancy widely used in memories.

Figure 9.20 shows (upper diagram) a row decoder that, in case of a defective row, can be replaced by a redundant row and its decoder set (lower diagram). The defective row can be disabled by burning a fuse, while the redundant decoder can be programmed by burning specific fuses. It must be noticed that the redundant decoder has a double number of inputs to accommodate all possible addresses of the defective row. The area occupancy attributed to the redundant modules must be taken into account when calculating the overall yield. In practice, it is possible to determine the optimal number of redundant blocks under certain hypotheses of defect density distribution. The first example of redundancy in a memory was in a 16kb chip designed at IBM. It contained 4 redundant worldlines and 6 redundant bitlines plus the relative decoders, with 7% area occupation.

The present memory chip dimensions make necessary the array partitioning into subarrays to reduce leakage currents and access times. As a consequence, an inefficient partitioning of the redundant rows and columns is realized. The use of

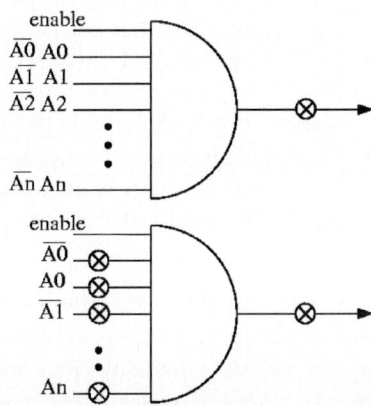

Fig. 9.20 Logic scheme of row decoders reconfigurable by means of fuses

globally redundant rows and columns would be more flexible and efficient, but with an unacceptable area overhead.

Other techniques use *error correcting codes* (ECCs). The use of ECCs allows a direct improvement in device reliability. The required area overhead is significant and, in addition, ECCs necessarily reduce access time.

A different approach uses an associative redundancy scheme, in which, instead of redundant rows and columns, redundant blocks are used to replace entire groups of cells. The addresses of a defective block are converted into addresses of the redundant block thanks to an associative memory storing the addresses of the defective blocks. This approach works correctly in the presence of defects that are not uniformly distributed. Access time to the associative memory slightly increases the total access time. The total area overhead is larger compared to the row/column redundancy scheme.

Today, new redundancy schemes are being explored, in which all these techniques are combined to achieve higher yields.

9.5.3 Burn-In and Screening

In the initial production phase of a new technology (often called rump-up phase), the number of defective devices is usually high. Burn-in techniques allow a successful detection of faulty devices before selling. It must be noticed that the burn-in phase, traditionally related to temperature tests, can also be regarded more generally as a further stress phase, in which the specific quantities of interest in the relevant failure mechanisms are investigated. A typical screening test in cellular phones is based on temperature cycles between -30 and $85\,°C$, together with power on/off and transmission cycles. It must be observed that the burn-in efficiency strongly depends on several factors:

- The burn-in duration must be optimal: a long burn-in can be dangerous since it causes an earlier component wearout; a short burn-in does not allow fully detecting the defective population.
- Burn-in stresses may influence the different failure mechanisms in different ways. The physical knowledge of the targeted failures is mandatory (it can be derived from HALT tests in phase 2).
- Constant failure rates do not improve with burn-in.
- The activation energy of the failure of defective population must be significantly higher compared to that of the normal population.

Figure 9.21 shows the impact of a burn-in on the failure rate of a system where several failure mechanisms are active. A 5-hour burn-in allows a significant reduction of the infant mortality phase without affecting the system wearout. A 20-hour burn-in allows a further infant mortality reduction, but it is already possible to

Fig. 9.21 Failure rate trend after burn-in of different durations

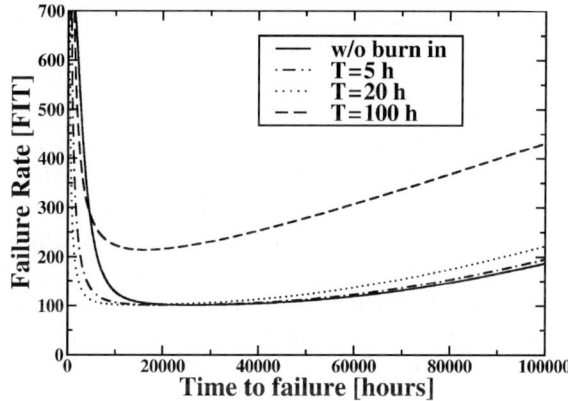

observe a higher failure rate in aged devices. Finally, a 100-hour burn-in has a totally negative effect both on useful life and wearout, without improving infant mortality.

References

1. Perera UD (1995) Reliability of mobile phones. Proceedings of Reliability and Maintainability Symp, pp. 33–38
2. Park SJ, Ha JS, Choi WS, Park SD (2002) Improving the reliability of a newly-designed mobile phone by environmental tests and accelerated life tests. Proceedings of Reliability and Maintainability Symp, pp. 642–646
3. Crowe D, Feinberg A (2001) Design for reliability. CRC Press
4. Ohring M (1998) Reliability and failure of electronic materials and devices. Academic Press
5. Koren I, Koren Z (1998) Defect tolerance in VLSI circuits: techniques and yield analysis. Proc IEEE, 86:1819–1837

Chapter 10
How to Read a Datasheet[1]

M. Dellutri, F. Lo Iacono and G. Campardo

10.1 Introduction

The following chapter will try to give a methodology and a guideline to read a datasheet of Multi-Chip. It contains so much information that sometimes it will not be immediately understandable. But once the main points of the datasheet have been explained, it will be easier to understand it. At the end of this chapter you will find an example of a datasheet, not for an existing product, but created on purpose to explain the different parts.

10.2 Contents of the First Page

The first page of the datasheet can be compared with the identity card of the device. At the top is placed the *commercial product* which is the name (normally it's placed on the top of the package). On the last page of the datasheet there is the *part numbering* scheme which explains the meaning of all the digits, in order to give the customer an accurate explanation of them. On the first page, below the product name, are shown the different types of memory that make up the Multi-Chip. The

M. Dellutri
Numonyx, Catania, Italy

F. Lo Iacono
Numonyx, Catania, Italy

G. Campardo
Numonyx, Agrate, Italy

[1] *Datasheet* is the word used to explain the set of specifications that fully describe the device. The datasheet is similar to a contract between supplier and customer. The DS of the memories to which we shall refer can be found at: www.st.com in Products/Technical Literature/Datasheets. The part numbers of each component are: Flash NOR M29WR128F, Flash NAND NAND128W4A, PSRAM M69AB128BB, and SRAM M68AR512D.

R. Micheloni et al. (eds.), *Memories in Wireless Systems,*
© Springer-Verlag Berlin Heidelberg 2008

memories which are included inside the Multi-Chip are: NOR Flash, NAND Flash, PseudoRAM (PSRAM), and Static RAM (SRAM). Of each single memory follows a brief description of main features and, moreover, there is a common part for all memories. Before describing in detail each single memory, it is important to understand the architectural differences on which depend the different characteristics of memory size, timing, access time, and data storage. The memories are divided into volatile and nonvolatile. The memories SRAM and PSRAM belong to the category of volatile memories, in the sense that all information stored is lost if the memory supply voltage is removed. On the other hand, the nonvolatile Flash memories with architecture NOR or NAND don't lose the information. With NOR architecture all cells are arranged in rows (wordline) and columns (bitline). The wordline is represented by all the control gates of the cells that belong to the same line. The bitline is represented by all drains that belong to the same column. The source line is the line where the sources of all cells belonging to the same sector are connected. This structure is typically made to 8, 16, or 32 bits, to agree with the number of input/outputs available from the memory on the data bus, organized, respectively, as byte, word, or double word. The NAND architecture is obtained by connecting 16 cells in series between a bitline and a source line. The most important advantages of this solution are the reduction of the array due to a lower number of contacts (each 16 cells) and the scaling of the junctions of source and drain. There is a different structure for the SRAM memory, whose cell is nothing more than a FlipFlop set-reset type, so a single bit is composed of 6 transistors; this structure allows for fast memory access, but requires a greater area of the silicon. PSRAM is used for a large memory capacity because, having a memory cell consisting of a pass transistor and a capacitor, it requires an area of silicon much smaller than in the case of SRAM [1,2]. So we note that:

- The NOR Flash has fast access time (90 ns) compared to NAND Flash (12 μs), but once all the contents of the page have been transferred into its internal buffer, it can read data sequentially with access times (for a page) of 50 ns.
- The NAND Flash has an address/data bus multiplexed (so a lower pin numbers to bring out), unlike NOR Flash.
- The two Flash memory types have different organization arrays: the NOR is organized into four banks (16Mb + 48Mb + 48Mb + 16Mb) with the parameters block placed both at the top and at the bottom of the array, while the NAND is organized in blocks, each block containing 32 pages and each page made up of 264 words.
- The two Flashes moreover have a number of different characteristics depending on the applications that will be used.

The differences between PSRAM and SRAM, at the datasheet level, are not as evident as in the case shown above, but we can observe that:

- SRAM has an access time more or less equal to PSRAM.
- PSRAM supports synchronous burst read and write and the "partial array self refresh" that allows restricting the refresh to only a portion of the array, with the advantage of reducing consumption.[2]
- To minimize consumption, SRAM supports "data retention" which allows keeping the data for a specific period of time with a power supply voltage well below the minimum value. From these differences we can guess the advantage of using a Multi-Chip, which can integrate NOR Flash, used for "code storage;" NAND Flash, for "mass storage;" PSRAM as high-capacity XiP memory; and SRAM as cache memory for storage backup during the downloading of data, or when the device is in standby mode. On the first page of the datasheet, we find a section common to all the memories. This normally shows the voltage range of Multi-Chip which, in this case, goes from 1.7 to 1.95 V. This range couldn't match with the standalone range; for example, the NOR Flash memory could have an extended voltage range. In order to ensure the functioning of Multi-Chip, the range is restricted in this way, so that all four memories are guaranteed to avoid over or under voltage that could damage or create bad functioning of any memory.

10.3 Ball Out and Functional Block Diagram

One of the most important steps in the design of a Multi-Chip is the studying of ball out, which as reported in the datasheet, is the allocation for each ball of signals as control, power-ground, address and data, which is necessary to ensure the full functionality of the Multi-Chip.[3] Usually the assignment of signals to the ball is provided by the customer, since he has developed the board on which the Multi-Chip will be assembled and is therefore aware of any board layout constraints. With reference to page 3 of the datasheet, we note that not all the balls have an unique assignment, but, on the contrary, either some balls are assigned different signals (e.g., E1, C6, C7, etc.), or the same signal is shared with more memories (e.g., C4, D6, F5, etc.). The need to share different signals of memories on a single ball stems from the need to minimize the number of balls; and this is reflected on the one hand by the possibility of reducing the size of the package, and on the other hand by reducing the number of tracks on board, all favoring a more compact system. In order to better understand the meaning of signals associated with various balls, the datasheet reports the signals' descriptions. In the section "signal descriptions" we can find, in addition to the description of the signal, whether it is shared with other

---

[2] The nonvolatile memories, SRAM and PSRAM, lose the information when the power supply is removed. The PSRAM needs further action to refresh. In fact, the SRAM is said to be static, because the information remains in memory until there is an available power supply. The PSRAM belongs to the family of dynamic RAM.

[3] The package is an FBGA: Fine Pitch Ball Grid Array. The ball out is an array of balls.

memories, and also which logical state relates to the type of operation that we want to perform. Below you will find a description of NOR Flash signals only, while for the remaining signals of the other memories you can refer to the datasheet reported at the end of the chapter. In detail:

- Address (A0–A22): The address inputs select the cells in the memory array; are in common with PSRAM.
- Data (DQ0–DQ15): the data I/O outputs give the data stored at the selected address during a bus read operation or input a command or the data to be programmed during a bus write operation. They are in common with PSRAM.
- Chip Enable (E_F): The chip enable input activates the memory control logic, input buffers, decoders, and sense amplifiers. When chip enable is low the device is in active mode. When chip enable is high the Flash memory is deselected, the outputs are high impedance, and the power consumption is reduced to the standby level.
- Output Enable ($G_{F/P}$): active low, the output enable pin controls data outputs during bus read operation; it's in common with PSRAM.
- Write Enable ($W_{F/P}$): active low, it controls the memory bus write operation. The data and address inputs are latched on the rising edge of chip enable; it's in common with PSRAM.
- Clock (K): The clock input synchronizes the memories for the microcontroller during synchronous read operations. It's not used when the memory works in asynchronous mode.
- Write Protect (WP_F): active low, it is an input that provides additional hardware protection for each block. When write protect is low, lock-down is enabled and any operation (program and erase) will be ignored.
- Reset/Block Temporary Unprotect (RP_F): it provides hardware reset of the memory or temporarily unprotects all blocks previously protected.
- Latch Enable ($L_{F/P}$): it latches the address bits on its rising edge; it's in common with PSRAM.
- Wait: it is used both during the synchronous operation, indicating when the data output are valid; and during asynchronous operation, acting as a signal ready/busy (when it is low it informs about the ongoing operation, while if it is high it indicates that the memory is ready to receive a new operation).
- Vpp_F: it is an additional power supply pin provided to the memory, allowing it to speed up program and erase operations.

The device also receives power supplies Vcc_F and Vss_F for the core, and $Vccq_F$ and $Vssq_F$ for the I/O buffers. Understanding all signals shown in the ball out, we can explain the functional block diagram on page 5 of the datasheet. There is a relationship between the ball out and the block diagram because once the ball out is fixed, which is the design constraint of the Multi-Chip in terms of signals, the block diagram is a direct consequence. This diagram is shown in the datasheet to give to the reader the visual idea of how the various memories of the Multi-Chip and their associated signals have been internally connected. The block diagram is divided into two lower blocks: the first is shared with NOR Flash and PSRAM, the second is shared with NAND Flash and SRAM. This subdivision is because of the different types of memories: NOR Flash is used as "code storage," while NAND Flash is used as "data storage" [3]. Each memory shares signals, whether with PSRAM or with SRAM. From the block diagram, we can see that the voltage supply for the "core" is separated for the different memories, in order to minimize disturbances when one or more memories are working simultaneously. In addition, the chip enable settings of various memories are separated. This choice is necessary if we want to use the memories separately. Considering the first block (NOR Flash and PSRAM): given that the two memories have addresses and data buses in common, it's not possible to simultaneously enable memories because this could create a conflict on the bus; so, except for some operations (which we will see later), it doesn't allow more than one memory to be active at the same time. This will only be possible if the two chips are separated. In fact, if the chip enable of first memory is a logical high and the second one is low, only the second memory can accept the input command and run all required operations while the first one will stay in stand by mode. Finally, we can say that, having two separate blocks with separated bus addresses, data, and all control signals, it is possible to work both the first block (NOR + PSRAM) and the second block (NAND + SRAM) in parallel in order to have a more compact and faster system.

10.4 Main Operations Allowed

The three basic operations that a nonvolatile Flash memory can perform are: read, program, and erase.

The purpose of the read operation is to determine the internal logic state of a cell, in other words, if the information inside the cell is 1 or 0 logic; whereas the program operation establishes the logic state of the cell depending on the data input. The erase operation puts all cells at logic state 1. In the case of the Multi-Chip the situation is more complicated, because the memories share the same data bus, and then the bus needs to know which kind of operations are permitted in order to avoid some operation on a memory which could cause a bus conflict. In the datasheet there is a section called "main operation mode" in which are shown all the operations allowed. In this specific case there are two tables, one on the block NOR and PSRAM and another on the block NAND and SRAM; this is because, as said before, neither block shares the data bus or the addresses, so they can be enabled simultaneously. On page 6 of the datasheet we can see that the table is organized with NOR Flash and PSRAM operations (read, write, outputs disable, standby) in the first column, while in the first row are the control signals, addresses, and data of both memories. For example, looking at the second row, to run a read operation of Flash, the control signals should be as follows: E_F#[4] (chip enable) and $G_{F/P}$# (output

---

[4] BAR signals are shown by the # symbol.

enable) low (V_{IL}); R_{PF} (reset), WP_F (write protect), $W_{F/P}$ (write enable) high (H_{IV}), since they are signals which have other functions; VPP_F "don' t care" (X), in other words its level can be either high or low. The pins A22–A0 are used to address the cells of memory, while on DQ0–DQ15 we could read the output data of memory. The signals of PSRAM must be set so that the memory is disabled, which means it is in standby or output disable mode. From the table we can note that when the Flash is in standby mode (the chip enable is high) some operations on PSRAM are possible, such as read and write. So looking at the table we can say that any operation on a memory that keeps the data bus busy cannot be done on another memory, because it would cause a conflict resulting in wrong data. In order to avoid this, we can find in the datasheet a sentence which explains that we can't enable the NOR Flash and PSRAM, or the NAND Flash and SRAM, at the same time. However, it is possible that in the near future, we will be able to perform some operations in parallel on both devices. The simultaneous operations will depend on the type of application in which the Multi-Chip will be used. Of course the simultaneous operations will give the customer the ability to get a faster system. For example, during an erase operation, which takes an execution time of hundreds of milliseconds, it will be possible to perform a write or read operation on another memory. Another parallel operation would be to read the Flash and, during the period that the data is valid, to enable a write operation to PSRAM so that the content of one goes into the other. In this case we can later access the data of the memory faster. These are some examples of operations in parallel, but before they can be put into the datasheet, it will be necessary for the chip maker to provide these operations in its testing flow in order to guarantee their reliability.

10.5 Maximum Ratings and DC-AC Parameters

Electronic devices require specific working conditions, first with reference to the electrical surroundings, but also considering temperature and humidity during the entire product life: from storing to final soldering onto the board. Looking at Table 5 in the datasheet, the first value is the allowed value of the temperature for product working conditions. The temperature has an impact not only on the leakage current and transistor threshold but also on the mechanical elements of the product, inducing deformation or functional failure. In the case of the Multi-Chip product, the working temperature range is the minimum of each range for each memory in order to guarantee all devices: for instance, the minimum temperature for a DRAM memory is higher than for a FLASH memory because of the junction leakage current that impacts the refresh frequency. The storage temperature is higher, assuming all memories are not biased. Biasing acts as an accelerator factor for temperature stress. The soldering temperature is not indicated, as the JEDEC specification is the agreed reference for the reflow temperature profile. As reviewed in Chap. 8, one of the major factors that induces failure on the board is exactly the reflow profile. The datasheet also provides information on the maximum voltage value that can be applied to I/Os and power supplies in order to en sure that the transistor gate oxide breakdown and latch-up phenomena are not triggered. Particularly interesting is the voltage called Voltage Identification, which allows a very high value, in our case 13.5 V. The explanation for this lies in the use of the "third level." Normally, input can be loaded with a signal that allows two logical values, corresponding to the voltage level: zero and VDD. If you want to produce a different logical level, particular buffers can be designed that are able to recognize, usually through a device that reads a particular threshold, a voltage value well above that of the power supply. In this case the buffer switches off the normal route and turns on a different one. These tricks are now used to enter the device into test mode or to access codes for identification of the device that are not in the addressable memory space. Note that the time of application of this potential is reduced to a second; this is because it is applied to a junction in the path test, a gate, in the normal way. The power VPPF may also be applied up to 13.5 V and is used to speed up the programming of the Flash, when it is programmed at the factory the first time. In this way the voltage is not internally generated using charge pumps, but applied externally, therefore with greater efficiency and reduced programming time. In this case, the potential is applied to a junction, so for a longer time. tVPPFH is granted time when the voltage is at the maximum value.

The short circuit current is calculated on a single output (a memory normally has 16) and its duration is for one second. That may seem short, but we have to remember that a 128 Mb memory, organized for 16 outputs, 8,388,608 words, with an access time of 100 ns, is read (all words sequentially) in 0.8388608 s! Finally, the power is dissipated: its value is for the SRAM, because this is the device with the higher power consumption, designed to be faster in the read operation. Again a consideration: suppose that we have a mobile phone with 20 devices inside and that all these devices have a consumption equal to an SRAM device, something like 20 W all together. Have you ever tried to put your fingers on a bulb filament of 25 W power consumption? Let's look at page 8 of the datasheet. The most important information, in this case, is not the power of the different devices, which will be commented on below, but the loads used for the measurements. We think that the devices are not designed to match the PCB application perfectly, as it is not known what exact impedances will load the pins of the device. The measure of access time, consumption, and all that, will be implemented by using a predefined network load that, a priori, is not the same for different devices. In Table 6 you can see how the values of resistors and capacitors change for different memories. The extreme case is that of PSRAM, where the load to the power is smaller than for the other devices.

This artifice, in the case of DRAM (PSRAM belongs to this family), helps a lot when switching from 0 logic state to 1 logic state, which is usually the slowest. The parallel of pull-up transistor with a small load greatly helps the transition that has a reference starting value of VDD/2, obtained powered the same pull-up resistance to VDD/2. Ultimately, in the case of the Multi-Chip, since the outputs of the different memories are shorted together, it is necessary to find a unique test which allows their characterization. The loads on page 8 is a reference for the customer, a point in space made up of temperature, load, and speed, useful in defining the PCB to be designed. Finally, Table 7 shows the output capacitance as viewed "watching" the pin, from the outside of the device. This capacitance results as the final contributions of package and transistors, joints, metallization, etc.

Because we have the junctions, we specify the conditions of measurement, the Test Condition, to take into account the effects of biasing. The allowed load conditions on the pin device is generally 30–50 pF in a Multi-Chip with 4 device outputs; since they are shorted together, we can have a higher value for the external load. The need to introduce impedance matching on the line will be one of the main topics of coming developments [4].

10.6 DC Parameters Benchmark Between Memories for the Multi-Chip

One of the key issues is the impact of consumption, which in specific applications, particularly mobile, is an aspect that should not be underestimated. In order to estimate the current consumption of the Multi-Chip the DC tables for each memory are reported in the datasheet. The most important parameters that are examined in a first analysis are these:

- supply current (program/erase);
- standby current.

Looking at pages 9 and 10 we see that the maximum power consumption for NOR Flash memory in read mode is 10 mA at 6 MHz asynchronous mode, while in burst mode it is 30 mA; it is 15 mA for NAND, while for PSRAM it is 30 mA and the SRAM is 12 mA. In standby mode the Multi-Chip offers considerable energy savings, because in this mode the NOR Flash has a maximum of 100 μA, NAND 50 μA, PSRAM 130 μA, and SRAM 20 μA.[5] Also from the DC tables we see that during program/erase operation the NOR Flash memory allows a current consumption of 20 mA and ensures 100,000 program/erase cycles; while the NAND Flash memory has a consumption of 15 mA, ensuring the same number of cycles as the NOR. Moreover, there are reported the values of input/output current leakage, referring to the single memory. In the case of the Multi-Chip we will need to calculate the total value (considering all four devices), simply by adding the contributions of each memory. Finally, the values of the thresholds VIH, VIL input and VOH, VOL output are reported; these values also are referred to a single device, and in the case of the Multi-Chip we will need to consider the restricted voltage range in order to have the values compatible for all four memories.

---

[5] This mode allows decreasing the access time. It is necessary to have an external clock signal to do different operations. During burst mode the memory stores reading in a time period and then begin to transfer data outside, every edge of clock, at the same time as new read operation.

10.7 How to Read Timing Chart and AC Table

The memory device communicates with other devices on the application, such as the baseband, through temporal sequences of input/output signals. The purpose of the following paragraphs is to provide a guideline to read the timing of the signals for the most important operations (program, erase, etc.) for the memory family assembled in the MultiChip.

10.8 NOR Memory Read Timings

In the section *Operating and AC Measurement Conditions* of the datasheet, Fig. 10.1, is shown the measurement conditions of characteristic parameters: the minimum and maximum values of power voltage, the minimum and maximum values of working temperature and load capacity, the characteristics of input signals such as rise and fall time, the voltage values on the high and low logical states, and finally what the threshold is (in our case it is fixed VCC/2) to determine when the signals are in low or high logic state.

Also we can see the schematic of the device under testing, including the capacitances between two power and ground voltages, and two resistances of 25 kΩ to bias the data bus, which puts the voltage at VCC/2 when the output buffer of the device is high impedance. All values are in their worst conditions, which are minimum voltage and maximum temperature. With reference to Fig. 10.2, the typical parameters regarding the read operation are: tAVAV, tAVQV, tELQV, tAVQV1, tELQV, tGLQV, tEHQZ, tGHQZ, tEHQX, tGHQX, tAXQX, and tWHGL. tAVAV, or *Address Valid to Address Valid*, is a cycle time for the reading of memory, and it is the minimum time gap which separates a successful read from the next one (Fig. 10.3). To perform a read operation in every matrix (in this case the cycle read is 250 ns), it's necessary to put the Chip Enable and Output Enable signals low and to disable the Write Enable, putting it high. The data is grabbed from 40 ns to 100 ns in steps of 2 ns. tAVQV, or *Address Valid to Data Valid*, is the time of read by address, that is the time in order that the data is output. Both Chip Enable and Output Enable have to be low. tELQV, or *Chip Enable Low to Data Valid*, is the read timing by Chip Enable when it's low, so the memory outputs from Standby and puts it in read mode. tGLQV, or *Output Enable Low to Data Valid*, is the access time by Output Enable. The measurement conditions expect the addresses are already stable and the Chip Enable is low. tEHQZ and tGHQZ, or *Chip Enable High to Output High-Z* and *Output Enable High to Output High-Z*, when one of the two events occurs, Chip Enable or Output Enable will be high, or in both conditions the output buffer of memory will be in high impedance.

After a minimum time equal to tEHQZ or tGHQZ, the electrical polarization data bus is managed from another device connected in parallel or at the voltage divider which forces VCC/2. tEHQX, tGHQX, or *Chip Enable, Output Enable to Output Transition* is the minimum delay between the rising edge of Chip Enable, Output

DC AND AC PARAMETERS

This section summarizes the operating measurement conditions, and the DC and AC characteristics of the device. The parameters in the DC and AC characteristics Tables that follow, are derived from tests performed under the Measurement Conditions summarized in Table 19., Operating and AC Measurement Conditions. Designers should check that the operating conditions in their circuit match the operating conditions when relying on the quoted parameters.

Table 19. Operating and AC Measurement Conditions

Parameter	M29WR128F, M29WR128FS				Unit
	70		90		
	Min	Max	Min	Max	
V_{CC} Supply Voltage	2.7	3.6	2.7	3.6	V
Ambient Operating Temperature	−40	85	−40	85	°C
Load Capacitance (C_L)	30		30		pF
Input Rise and Fall Times		10		10	ns
Input Pulse Voltages	0 to V_{CC}		0 to V_{CC}		V
Input and Output Timing Ref. Voltages	$V_{CC}/2$		$V_{CC}/2$		V

Figure 9. AC Measurement I/O Waveform

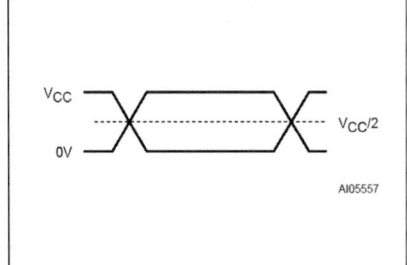

Figure 10. AC Measurement Load Circuit

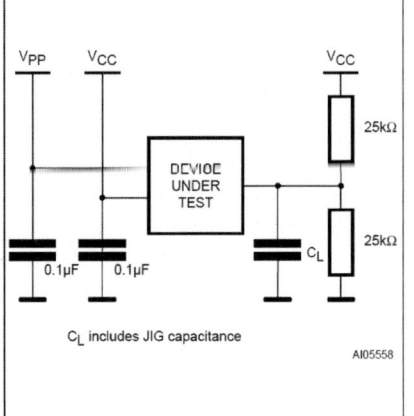

Table 20. Device Capacitance

Symbol	Parameter	Test Condition	Min	Max	Unit
C_{IN}	Input Capacitance	$V_{IN} = 0V$		6	pF
C_{OUT}	Output Capacitance	$V_{OUT} = 0V$		12	pF

Note: Sampled only, not 100% tested.

Fig. 10.1 Flash NOR, operating and AC measurement conditions

M29WR128F, M29WR128FS

Figure 11. Random Read AC Waveforms

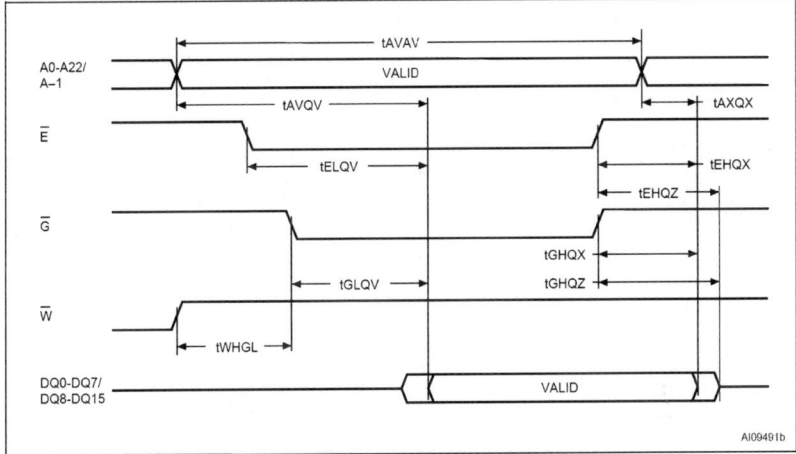

Note: 1. The Latch Enable signal, L̄, must be held Low, V$_{IL}$.
2. Bit CR19 of the Configuration Register is set to '1' (asynchronous mode).

Figure 12. Asynchronous Page Read AC Waveforms

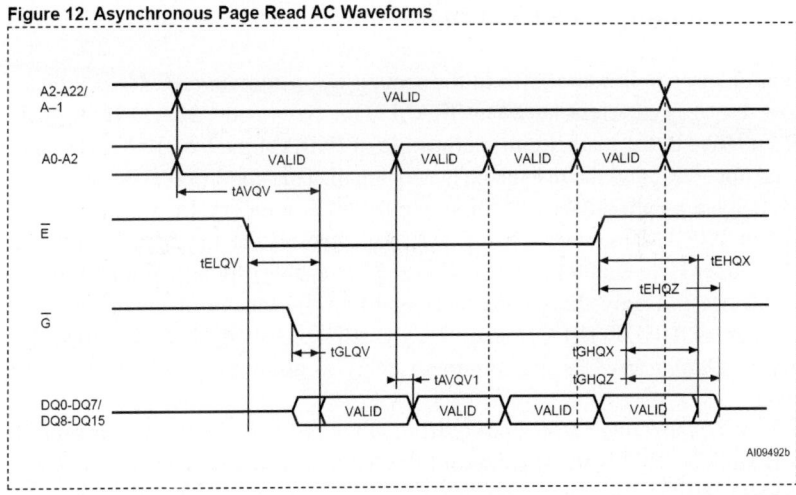

Fig. 10.2 NOR Flash memory signals for read mode

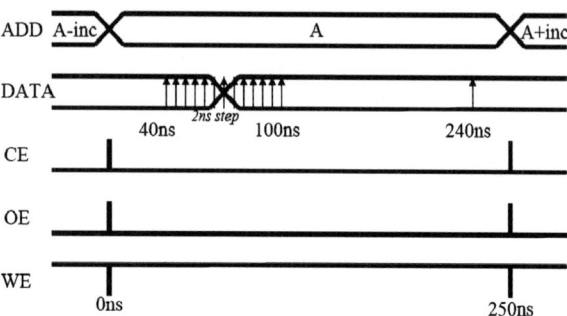

Fig. 10.3 tAVAV parameter measurement

Enable and the edge of the output signal. tWHGL, or *Write Enable High to Output Enable Low*, is the minimum time which can elapse between the rising edge of Write Enable and the falling edge of Output Enable, advising the memory to switch from command to read mode.

10.9 RAM Memory Read Timings

The measurement conditions shown in the section *Operating and AC Measurement Conditions* are very different between SRAM (Fig. 10.4) and PSRAM (Fig. 10.5) memories relative to the load circuit; in the case of PSRAM the load RC is biased at VCC/2 and not VCC, and the resistance is very small. This condition, inherited from SDRAM, helps the output data switching. Unlike NOR memory, the read operation of SRAM or PSRAM memories has the ability to select the upper or lower byte using UB (Upper Byte enable) and LB (Lower Byte enable) signals. With reference to Fig. 10.6, we put only the parameters regarding the enable of upper or lower data bus, such as tBHQZ, tBLQV, and tBLQX. tBHQZ, or *Upper/Lower Byte Enable High to Output Hi-Z*, is the minimum delay to allow the memory buffer to output data.

tBLQV, or *Upper/Lower Byte Enable High to Output Valid*, is the access time of UB or LB signals. This is the maximum time between the fall edge of UB or LB and data valid. tBLQX, or *Upper/Lower Byte Enable Low to Output transition*, is the minimum time delay between fall edge of UB and LB and the edge of output. Please note that it does not refer to data valid, at the beginning of outputs commutation that they will stabilize after a period equal tBLQV.

10.10 Read Timings for NAND Memory

In the case of NAND memory (Fig. 10.7), in order to make the migration toward high density easier without losing compatibility in terms of ball-out, the addresses and data output are multiplexed using only one I/O bus.

M68AR512D

DC AND AC PARAMETERS

This section summarizes the operating and measurement conditions, as well as the DC and AC characteristics of the device. The parameters in the following DC and AC Characteristic tables are derived from tests performed under the Measurement Conditions listed in the relevant tables. Designers should check that the operating conditions in their projects match the measurement conditions when using the quoted parameters.

Table 4. Operating and AC Measurement Conditions

Parameter		M68AR512D		Unit
		55	70	
V_{CC} Supply Voltage		1.65 to 1.95	1.65 to 1.95	V
Ambient Operating Temperature	Range 1	0 to 70	0 to 70	°C
	Range 6	−40 to 85	−40 to 85	°C
Load Capacitance (C_L)		30	30	pF
Output Circuit Protection Resistance (R_1)		15.3	15.3	kΩ
Load Resistance (R_2)		11.3	11.3	kΩ
Input Rise and Fall Times		1	1	ns/V
Input Pulse Voltages		0 to V_{CC}	0 to V_{CC}	V
Input and Output Timing Ref. Voltages		$V_{CC}/2$	$V_{CC}/2$	V
Output Transition Timing Ref. Voltages		$V_{RL} = 0.3V_{CC}$; $V_{RH} = 0.7V_{CC}$	$V_{RL} = 0.3V_{CC}$; $V_{RH} = 0.7V_{CC}$	V

Figure 5. AC Measurement I/O Waveform

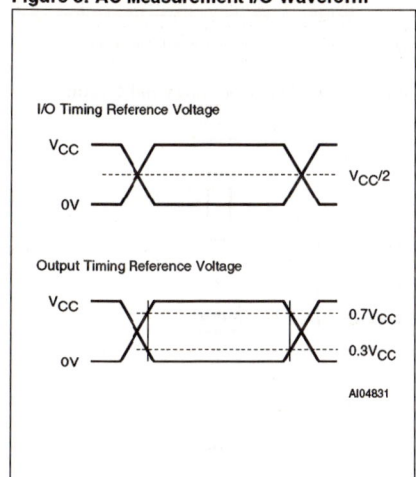

Figure 6. AC Measurement Load Circuit

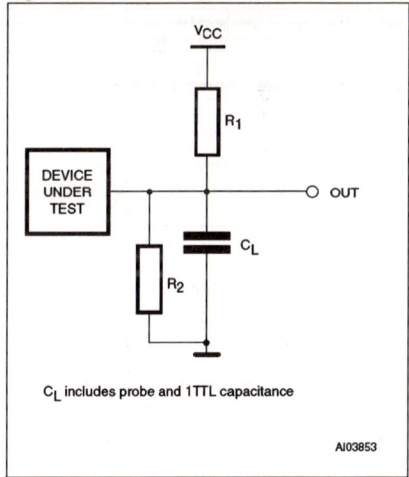

Fig. 10.4 Static RAM (SRAM), operating and AC measurement conditions

tBLEH, or *Upper/Lower Byte Enable Low to Chip Enable*, is the minimum delay between the falling edge of UB or LB and the rising edge of Chip Enable.

tBLWH, or *Upper/Lower Byte Enable Low to Write Enable High*, is the minimum delay between the falling edge of UB or LB and the rising edge of Write Enable.

tDVBH, or *Output Valid to Upper/Lower Byte Enable High*, is the minimum delay between data valid and the rising edge of UB or LB.

tELBH, or *Chip Enable to Upper/Lower Byte Enable High*, is the minimum delay between the falling edge of Chip Enable and the rising edge of UB or LB.

tWLBH, or *Write Enable Low to Upper/Lower Byte Enable High*, is the minimum delay between the falling edge of Write Enable and the rising edge of UB or LB.

10.13 NAND Memory Timings for Modify Operations

In the case of NAND memory (Fig. 10.12), typical operations are page programming and block erase. Command input protocol requires the use of AL (address latch) and CL (command latch) signals because the architecture used is multiplexing. The portion of bus used, whether x8 or x16 version, is always I/O0...I/O7. The internal finite state machine takes into account any command on the rising edge of the W signal or Write Enable when CL is kept high and AL low. The protocol provides for two or three cycles of command to ensure greater safety in the operation and to prevent unintended changes by the user. The main timer parameters for command input are: tCLHWL, tWHCLL, tWLWH, tDVWH, tWHDX, and tCLHWL.

tCLHWL, or *Command Latch High to Write Enable Low*, is the setup time, i.e., the minimum time distance between the rising edge of the CL signal and the falling edge of the W signal. The minimum value guaranteed on the datasheet is 0 ns.

tWHCLL, or *Write Enable High to Command Latch Low*, is the hold time, i.e., the necessary delay between the rising edge of the W signal and the falling edge of the CL signal. The minimum value guaranteed on the datasheet is 0 ns.

tWLWH, or *Write Enable Low to Write Enable High*, is the minimum width of the W signal pulse. The minimum value guaranteed on the datasheet is 25 ns when the supply voltage is 3.0 V, and 40 ns when the supply voltage is 1.8 V.

tDVWH, or *Output Valid to Write Enable High*, is the data setup time, i.e., how long data must be ready at the I/O bus before the rising edge of the W signal transfers the command to the internal finite state machine. The minimum value guaranteed on the datasheet is 10 ns.

tWHDX, or *Write Enable High to Output Transition*, is the data hold time, i.e., how long time should be kept stable data input on bus I/O referred to the rising edge of the W signal.

Figure 24. Command Latch AC Waveforms

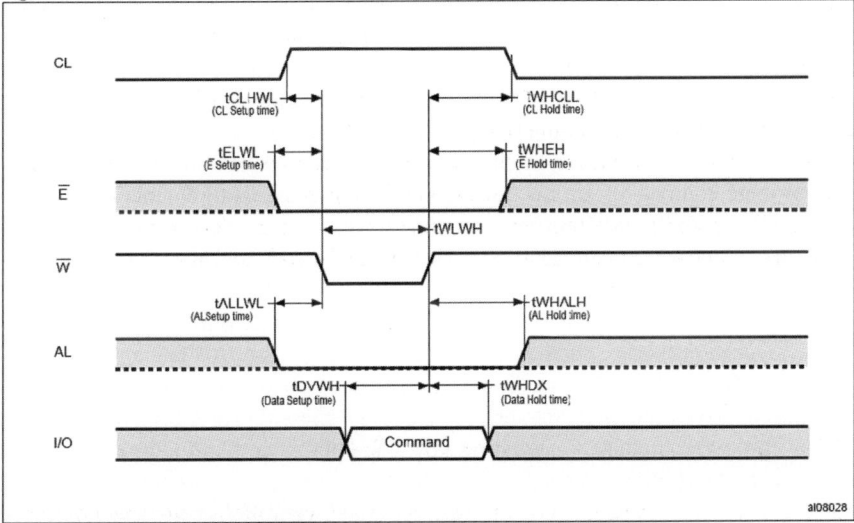

Figure 25. Address Latch AC Waveforms

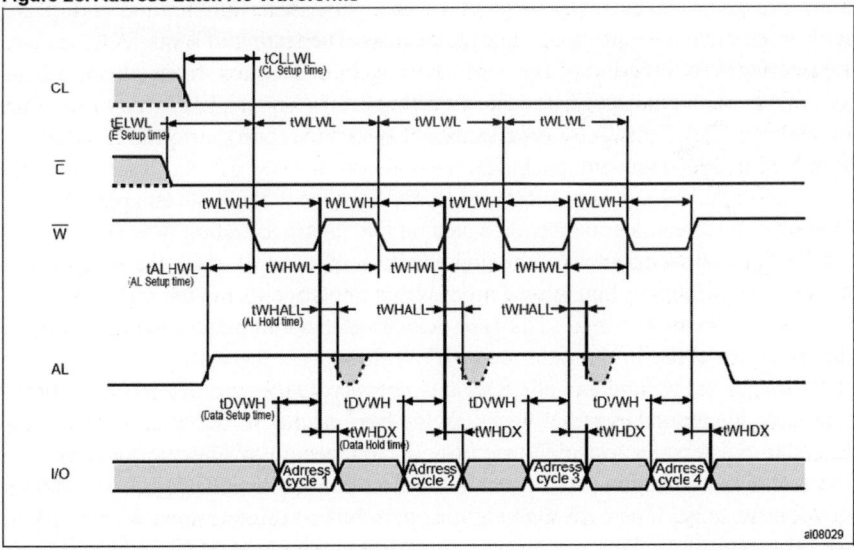

Note: Address cycle 4 is only required for 512Mb and 1Gb devices.

Fig. 10.12 NAND Flash memory signals for modify operations

10.14 From Timing to Testing

The MultiChip product is a real system on a package, which is a complex system that is defined in the assembly phase. Each device is connected internally through a common substrate in order to operate independently of the others, and only one device at a time can be selected and use both addresses and data bus while others are deselected. The check of the functionality of the MultiChip product is therefore submitted to the correct operation of each component that set the system. It's sufficient that only one component not be good to compromise the requirements in terms of functionality of the MultiChip. The electrical testing of each device is done through two types of measurement: the DC measurement, which verifies aspects such as the current consumption or the minimum and maximum logical voltage levels; and dynamic measures, or AC, that verifies the timing as reported into the datasheet in various operation modes, such as read, program, and erase. All DC and AC measures typically used into testing flow is very important tool whether to validate in the phase of development of the device or quality guarantee of the product in high volumes production phase. The testing flow can be applied to an individual device or following the application conditions by the datasheet; and in this case we speak in terms of test sequence user-mod, Un-like test-mode where following special access protocols, not available on the datasheet, is possible to make observable internal circuit blocks at the memory or to make accessible the configuration register in order to optimize the performance of the devices. The testing of a single die is done in three stages of tests during the production cycle of memory, from silicon wafers to the final production package. The three types of tests are EWS (performed on devices on wafer), Final Test, and Electrical Characterization (performed on devices assembled in the production package).

The acronym EWS means Electrical Wafer Sort and is the first test applied at the wafer level to screen a good device based on the electrical testing flow. The testing flow EWS is a flow oriented to maximize the electrical yield of the wafer, typically affected by the defects of the spread process that impacts either in the matrix or in the circuitry. The main aim is to search for devices with a complete matrix, or devices with defective cells. In the latter case, EWS flow gives the ability to recover lost bits by means of redundant cells. Devices not recoverable are discarded. In order to activate bit redundancy, it's necessary to have access to the matrix for setting particular registers; the test-mode memory access during EWS testing is much more powerful than user-mode memory access. Before the wafer sawing process, devices not good are subject to an inked phase; this permits us to isolate good dice from bad dice. The typical thickness of a wafer, for a standalone assembly, is 240 μm; this thickness reduction process is obtained through a mechanical-chemical treatment called lapping.

When there are several devices to be assembled (as in the MultiChip case), the thickness of any die is less than the standalone case (the typical thickness is 100 μm). In this case it's better to avoid the inked phase because of possible reliability problems during the lapping and assembly process. It's preferable to adopt another procedure, called inkless, utilizing (x,y) position coordinates of good dice contained in a map. The EWS testing phase is applied to all memory types treated in this chapter: flash NOR/NAND and RAM (type SRAM and PSRAM). Different dice with various kinds of memories are thus assembled in a single MultiChip package, forming the final product. For the purpose of optimizing the total yield of the MultiChip, EWS testing is integrated with different tests oriented to user-mode functionality verification (like the speed sorting test, where the speed class is verified in order to guarantee the datasheet specifications). The EWS flow so modified is now named KGD, i.e., Know Good Die. After assembling, the MultiChip products are subject to the Final Test. The Final Test is oriented to verify the proper operation of every single component assembly, after verifying the correctness of the electrical connections inside the package.

The Final Test also ensures compliance with all DC and AC specifications on the datasheet. The MultiChip flow test is performed, composing each stream in sequence on each device inside the package. In some cases, to increase productivity, it exploits the ability to run in parallel operations for the cancellation of NAND/NOR memories and the writing/reading for the memories and PSRAM/ SRAM. A sample of MultiChip that has exceeded the Final Test, typically 10 pieces for three batches spread to a total of 30 pieces, is under full electrical characterization, where all parameters on the datasheet are measured and evaluated in terms of performance, according to all aspects related to variations in temperature, supply voltage, and load. During the development phase of a product MultiChip, the engineering team aims to maximize yields in EWS and Final Tests in order to minimize the economic loss due to discarded devices; after assembly, it becomes a financial loss related to costs associated with the package. The work of yield enhancement is based on the correlation between the two streams test EWS and the Final Test, in order to cover as many tests as possible, applying the necessary margins based on the results of Electrical Characterization.

References

1. Campardo G, Micheloni R, Novosel D (2005) VLSI-design of nonvolatile memories. Springer Series in Advanced Microelectronics
2. Cappelletti P, Golla C, Olivo P, Zanoni E (1999) Flash memories. Kluwer Academic Publishers, Norwell, MA
3. Technical Article. An unprecedented offer of high density flash memories for code and data storage. www.st.com
4. Technical Note. Timing specification derating for high capacitance output loading. www.micron.com

Figure 3. LFBGA Connections (Top view through package)

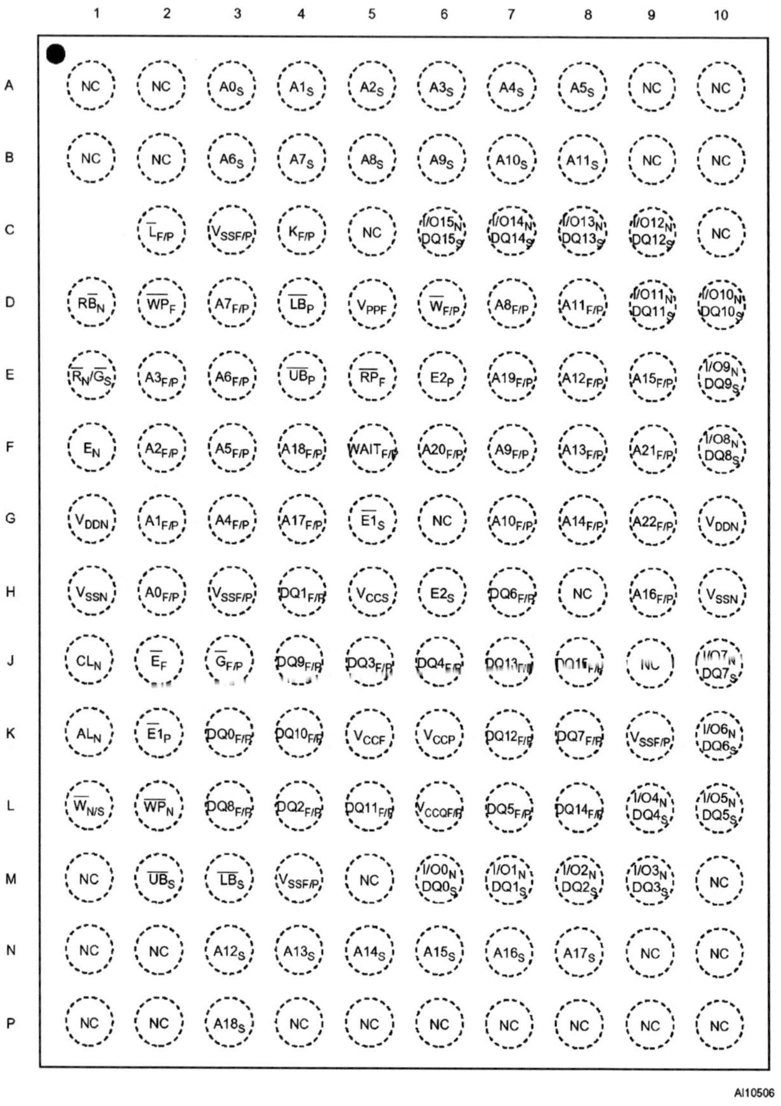

	1	2	3	4	5	6	7	8	9	10
A	NC	NC	$A0_S$	$A1_S$	$A2_S$	$A3_S$	$A4_S$	$A5_S$	NC	NC
B	NC	NC	$A6_S$	$A7_S$	$A8_S$	$A9_S$	$A10_S$	$A11_S$	NC	NC
C		$\overline{L}_{F/P}$	$V_{SSF/P}$	$K_{F/P}$	NC	$I/O15_N / DQ15_S$	$I/O14_N / DQ14_S$	$I/O13_N / DQ13_S$	$I/O12_N / DQ12_S$	NC
D	$\overline{RB}_N$	$\overline{WP}_F$	$A7_{F/P}$	$\overline{LB}_P$	V_{PPF}	$\overline{W}_{F/P}$	$A8_{F/P}$	$A11_{F/P}$	$I/O11_N / DQ11_S$	$I/O10_N / DQ10_S$
E	$\overline{R}_N/\overline{G}_S$	$A3_{F/P}$	$A6_{F/P}$	$\overline{UB}_P$	$\overline{RP}_F$	$E2_P$	$A19_{F/P}$	$A12_{F/P}$	$A15_{F/P}$	$I/O9_N / DQ9_S$
F	E_N	$A2_{F/P}$	$A5_{F/P}$	$A18_{F/P}$	$WAIT_{F/P}$	$A20_{F/P}$	$A9_{F/P}$	$A13_{F/P}$	$A21_{F/P}$	$I/O8_N / DQ8_S$
G	V_{DDN}	$A1_{F/P}$	$A4_{F/P}$	$A17_{F/P}$	$\overline{E}1_S$	NC	$A10_{F/P}$	$A14_{F/P}$	$A22_{F/P}$	V_{DDN}
H	V_{SSN}	$A0_{F/P}$	$V_{SSF/P}$	$DQ1_{F/P}$	V_{CCS}	$E2_S$	$DQ6_{F/P}$	NC	$A16_{F/P}$	V_{SSN}
J	CL_N	$\overline{E}_F$	$\overline{G}_{F/P}$	$DQ9_{F/P}$	$DQ3_{F/P}$	$DQ4_{F/P}$	$DQ13_{F/P}$	$DQ15_{F/P}$	NC	$I/O7_N / DQ7_S$
K	AL_N	$\overline{E}1_P$	$DQ0_{F/P}$	$DQ10_{F/P}$	V_{CCF}	V_{CCP}	$DQ12_{F/P}$	$DQ7_{F/P}$	$V_{SSF/P}$	$I/O6_N / DQ6_S$
L	$\overline{W}_{N/S}$	$\overline{WP}_N$	$DQ8_{F/P}$	$DQ2_{F/P}$	$DQ11_{F/P}$	$V_{CCQF/P}$	$DQ5_{F/P}$	$DQ14_{F/P}$	$I/O4_N / DQ4_S$	$I/O5_N / DQ5_S$
M	NC	$\overline{UB}_S$	$\overline{LB}_S$	$V_{SSF/P}$	NC	$I/O0_N / DQ0_S$	$I/O1_N / DQ1_S$	$I/O2_N / DQ2_S$	$I/O3_N / DQ3_S$	NC
N	NC	NC	$A12_S$	$A13_S$	$A14_S$	$A15_S$	$A16_S$	$A17_S$	NC	NC
P	NC	NC	$A18_S$	NC	NC	NC	NC	NC	NC	NC

AI10506

SIGNAL DESCRIPTIONS

See Figure 2., Logic Diagram and Table 1., Signal Names, for a brief overview of the signals connected to this device.

Address Inputs (A0-A22). Addresses A0-A18 are common inputs for the NOR Flash memory, SRAM and PSRAM components. The other lines (A19-A22) are inputs for the NOR Flash memory and PSRAM components only.

The Address Inputs select the cells in the memory array to access during Bus Read operations in the NOR Flash memory and also during Write operations in the PSRAM and SRAM.

During NOR Flash memory Bus Write operations they control the commands sent to the Command Interface of the Program/Erase Controller.

In the PSRAM these signals are also used during the Set Configuration Register sequence.

Inputs/Outputs (I/O0/DQ0-I/O7/DQ15). I/O0/DQ0-I/O7/DQ15 are used by the NAND Flash memory and the SRAM components.

In the NAND Flash, the Input/Outputs are used to input the selected address, output the data during a Read operation or input a command or data during a Write operation. Command and Address Inputs only require I/O0 to I/O7.

In the NAND Flash, the inputs are latched on the rising edge of Write Enable. I/O0-I/O15 are left floating when the device is deselected or the outputs are disabled.

In the SRAM the Upper Byte Data Inputs/Outputs carry the data to or from the upper part of the selected address during a Write or Read operation, when Upper Byte Enable (UB) is driven Low.

The Lower Byte Data Inputs/Outputs carry the data to or from the lower part of the selected address during a Write or Read operation, when Lower Byte Enable (LB) is driven Low.

Data Inputs/Outputs (DQ0-DQ15). The Data I/O are used by the NOR Flash memory and PSRAM components only.

The Data I/O output the data stored at the selected address during a Bus Read operation in the NOR Flash memory, and also during Bus Write operations in the PSRAM.

During NOR Flash memory Bus Write operations they represent the commands sent to the Command Interface of the Program/Erase Controller.

NOR Flash Memory Chip Enable ($\overline{E_F}$). The Chip Enable pin, $\overline{E_F}$, activates the NOR Flash memory, allowing Bus Read and Bus Write operations to be performed. When Chip Enable is High, V_{IH}, all other pins are ignored.

It is not allowed to have $\overline{E_F}$ at V_{IL}, $\overline{E_N}$ at V_{IL}, $\overline{E1_P}$ at V_{IL}, $E1_S$ at V_{IL}, $E2_P$ at V_{IH} and $E2_S$ at V_{IH} at the same time. Only one memory component can be enabled at a time.

NOR Flash memory/PSRAM Output Enable ($\overline{G_{F/P}}$). The Output Enable pin, $\overline{G_{F/P}}$, is common to the NOR Flash memory and PSRAM components.

It controls the Bus Read operation of the memory.

NOR Flash memory/PSRAM Write Enable ($\overline{W_{F/P}}$). The Write Enable pin, $\overline{W_{F/P}}$, is common to the NOR Flash memory and PSRAM components.

It controls the Bus Write operation of the memory.

NOR Flash memory/PSRAM Clock ($K_{F/P}$). The Clock, $K_{F/P}$, is common to the NOR Flash memory and PSRAM components.

See the M29WR128F and M69AB128BB datasheets for details.

NOR Flash memory Write Protect ($\overline{WP_F}$). The Write Protect pin provides a hardware method of protecting the eight outermost parameter blocks (four at the top, and four at the bottom of the address space).

When $\overline{WP_F}$ is Low, V_{IL}, the memory protects the eight outermost parameter blocks; program and erase operations in these blocks are ignored while $\overline{WP_F}$ is Low, even when $\overline{RP_F}$ is at V_{ID}.

When $\overline{WP_F}$ is High, V_{IH}, the memory reverts to the previous protection state of the eight outermost parameter blocks (four at the top, and four at the bottom of the address space). Program and erase operations can now modify the data in these blocks unless the blocks are protected using Block Protection.

NOR Flash memory Reset/Block Temporary Unprotect ($\overline{RP_F}$). The Reset/Block Temporary Unprotect pin can be used to apply a Hardware Reset to the memory or to temporarily unprotect all the blocks previously protected using a High Voltage Block Protection technique (In-System or Programmer technique).

Note that if $\overline{WP_F}$ is at V_{IL}, then the eight outermost parameter blocks will remain protected even if $\overline{RP_F}$ is at V_{ID}.

A Hardware Reset is achieved by holding Reset/Block Temporary Unprotect Low, V_{IL}, for at least t_{PLPX}. After Reset/Block Temporary Unprotect goes High, V_{IH}, the memory will be ready for Bus Read and Bus Write operations after t_{PHEL} or t_{RHEL}, whichever occurs last.

Holding $\overline{RP_F}$ at V_{ID} will temporarily unprotect the blocks protected by a High Voltage Protection technique. Program and erase operations on all blocks will be possible. The transition from V_{IH} to V_{ID} must be slower than t_{PHPHH}.

FUNCTIONAL DESCRIPTION

The NAND Flash, PSRAM, SRAM and NOR Flash memory components have separate power supplies but the NAND Flash and the SRAM on the one hand, and the NOR Flash memory and the PSRAM on the second hand, share the same grounds. They are distinguished by six Chip Enable inputs: E_F for the NOR Flash memory, E_N for the NAND Flash, $E1_P$ and $E2_P$ for the PSRAM, and $E1_S$ and $E2_S$ for the SRAM.

Recommended operating conditions do not allow the NOR Flash memory and PSRAM components, or the NAND Flash memory and SRAM components to be active at the same time. The most common example is simultaneous read operations on the NOR Flash memory and the PSRAM which would result in a data bus contention. Therefore it is recommended to put the other devices in the high impedance state when reading the selected device.

Figure 4. Functional Block Diagram

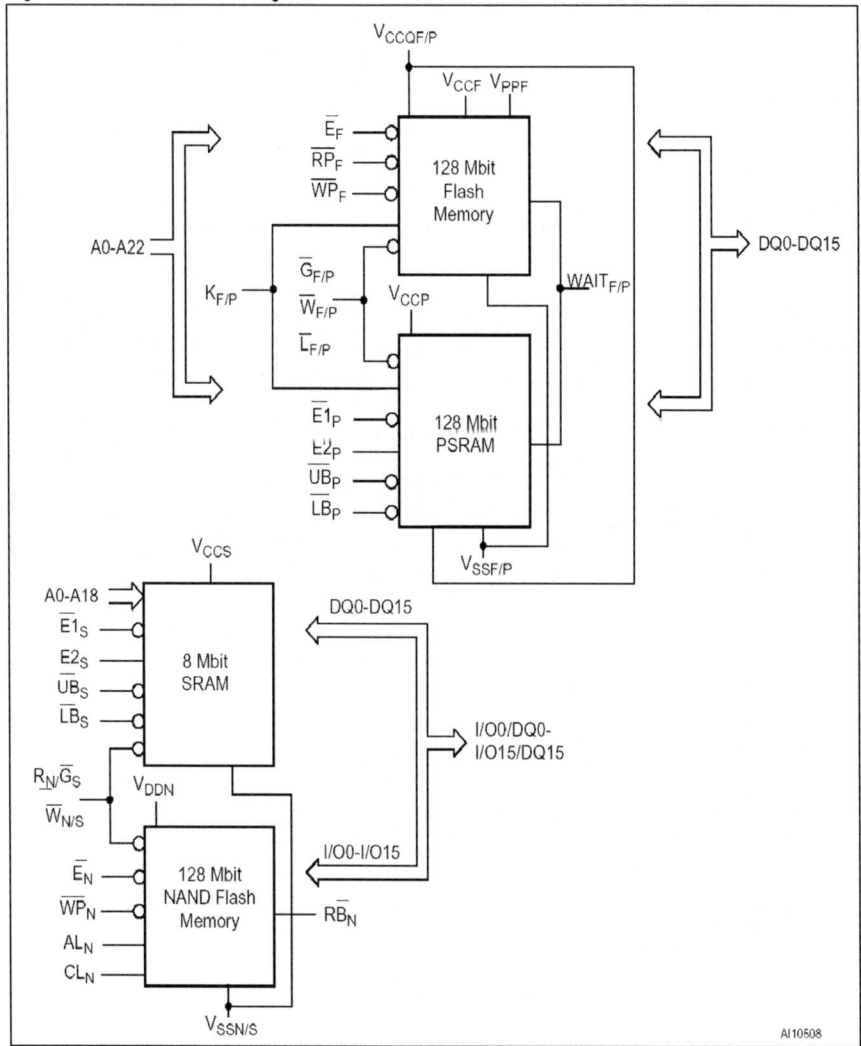

Table 3. NOR Flash memory and PSRAM Operating Modes– PSRAM in Asynchronous Mode

Operation	$\bar{E}_F$	$\overline{RP}_F$	$\overline{WP}_F$	V_{PPF}	$\overline{G}_{F/P}$	$\overline{W}_{F/P}$	$E2_P$	$\overline{E1}_P$	K_P	$\overline{L}_P$	$\overline{LB}_P$	$\overline{UB}_P$	$\overline{WAIT}_P$	A22-A0	DQ0-DQ7	DQ8-DQ15
NOR Flash Bus Read	V_{IL}	V_{IH}	V_{IH}	XX	V_{IL}	V_{IH}	The PSRAM should be disabled							Cell Address	Data Output	
NOR Flash Bus Write	V_{IL}	V_{IH}	V_{IH}	$>V_{IH}$	V_{IH}	V_{IL}								Command Address	Data Input	
NOR Flash Output Disable	X	V_{IH}	V_{IH}	XX	V_{IH}	V_{IH}	Any PSRAM mode is allowed							X	Hi-Z	
NOR Flash Standby	V_{IH}	V_{IH}	V_{IH}	XX	X	X								X	Hi-Z	
PSRAM Standby (Deselected)	Any NOR Flash memory mode is allowed.				X	X	V_{IH}	V_{IH}	X	X	X	X	High-Z	X	High-Z	High-Z
PSRAM Output Disable[1]					V_{IH}	V_{IH}	V_{IH}	V_{IL}	X	(2)	X	X	High-Z	(3)	High-Z	High-Z
PSRAM Output Disable (No Read)					V_{IL}	V_{IH}	V_{IH}	V_{IL}	X	(2)	V_{IH}	V_{IH}	High-Z	Valid	High-Z	High-Z
PSRAM Power-Down[6]					X	X	V_{IL}	X	X	X	X	X	High-Z	X	High-Z	High-Z
PSRAM No Write					V_{IH}[5]	V_{IL}	V_{IH}	V_{IL}	X	(2)	V_{IH}	V_{IH}	High-Z	Valid	Invalid	Invalid
PSRAM Read	The NOR Flash memory should be disabled.				V_{IL}	V_{IH}	V_{IH}	V_{IL}	X	(2)	V_{IH}	V_{IL}	High-Z	Valid	High-Z	Output Valid
					V_{IL}	V_{IH}	V_{IH}	V_{IL}	X	(2)	V_{IL}	V_{IH}	High-Z	Valid	Output Valid	High-Z
					V_{IL}	V_{IH}	V_{IH}	V_{IL}	X	(2)	V_{IL}	V_{IL}	High-Z	Valid	Output Valid	Output Valid
PSRAM Page Read					V_{IL}	V_{IH}	V_{IH}	V_{IL}	X	(2)	V_{IL}/V_{IH}	V_{IL}/V_{IH}	High-Z	Valid	(4)	(4)
PSRAM Write					V_{IH}[5]	V_{IL}	V_{IH}	V_{IL}	X	(2)	V_{IH}	V_{IL}	High-Z	Valid	Invalid	Input Valid
					V_{IH}[5]	V_{IL}	V_{IH}	V_{IL}	X	(2)	V_{IL}	V_{IH}	High-Z	Valid	Input Valid	Invalid
					V_{IH}[5]	V_{IL}	V_{IH}	V_{IL}	X	(2)	V_{IL}	V_{IL}	High-Z	Valid	Input Valid	Input Valid

Note: 1. The device must not be kept in the Output Disable state during more than 1µs.
2. The address latch is transparent when $\bar{L}$ is at V_{IL}. The addresses are latched on the rising edge of $\bar{L}$ (except for A0-A2 during Asynchronous Page Read operations).
3. Address bits are either at V_{IL} or V_{IH} but must be valid before Read or Write operation.
4. Data outputs are either valid or Hi-Z depending on the state of LB/UB inputs.
5. $\bar{G}$ can be V_{IL} during a Write operation if the following conditions are satisfied:
   a. Write pulse is initiated by $\overline{E1}$, or cycle time of the previous operation cycle is satisfied;
   b. $\bar{G}$ stays V_{IL} during the entire Write cycle.
6. Power-Down mode can be entered from Standby state and all DQ pins are in High-Z state. Data retention depends on the Power-Down mode selected.
7. X = V_{IL} or V_{IH}, XX = V_{IL}, V_{IH}.

MAXIMUM RATING

Stressing the device above the rating listed in the Absolute Maximum Ratings table may cause permanent damage to the device. These are stress ratings only and operation of the device at these or any other conditions above those indicated in the Operating sections of this specification is not im-

plied. Exposure to Absolute Maximum Rating conditions for extended periods may affect device reliability. Refer also to the STMicroelectronics SURE Program and other relevant quality documents.

Table 5. Absolute Maximum Ratings

Symbol	Parameter	Value		Unit
		Min	Max	
T_{BIAS}	Temperature Under Bias	−25	85	°C
T_{STG}	Storage Temperature	−55	125	°C
T_{LEAD}	Lead Temperature during Soldering		(1)	°C
V_{IO}	Input or Output Voltage	−0.5	2.45	V
V_{CCF}	NOR Flash memory Supply Voltage	−0.6	4	V
V_{CCS}	SRAM Supply Voltage	−0.5	2.5	V
V_{CCP}	PSRAM Supply Voltage	−0.5	2.6	V
V_{DDN}	NAND Flash Supply Voltage	−0.6	2.7	V
$V_{CCQF/P}$	NOR Flash memory/PSRAM Input/ Output Supply Voltage	−0.6	4	V
V_{ID}	Identification Voltage	−0.6	13.5	V
$V_{PPF}^{(2)}$	Flash Program Voltage	−0.6	13.5	V
$I_O^{(3)}$	Output Short Circuit Current		20	mA
t_{VPPFH}	Time for V_{PPF} at V_{PPFH}		100	hours
P_D	SRAM Power Dissipation	1		W

Note: 1. Compliant with the JEDEC Std J-STD-020B (for small body, Sn-Pb or Pb assembly), the ST ECOPACK® 7191395 specification, and the European directive on Restrictions on Hazardous Substances (RoHS) 2002/95/EU.
2. V_{PPF} must not remain at 12V for more than a total of 80hrs.
3. One output at a time, not to exceed 1 second duration.
4. For more details refer to the respective datasheets.

DC AND AC PARAMETERS

This section summarizes the operating measurement conditions, and the DC and AC characteristics of the device. The parameters in the DC and AC characteristics Tables that follow, are derived from tests performed under the Measurement Conditions summarized in Table 6., Operating and AC Measurement Conditions. Designers should check that the operating conditions in their circuit match the operating conditions when relying on the quoted parameters.

Table 6. Operating and AC Measurement Conditions

Parameter	NAND Flash		SRAM		Flash Memories		PSRAM		Unit
	Min	Max	Min	Max	Min	Max	Min	Max	
V_{DDN} Supply Voltage	1.7	1.95	–	–	–	–	–	–	V
V_{CCS} Supply Voltage	–	–	1.7	1.95	–	–	1.7	1.95	V
V_{CCF} Supply Voltage	–	–	–	–	1.7	1.95	–	–	V
V_{CCP} Supply Voltage	–	–	–	–	–	–	1.7	1.95	V
$V_{DDQF/P}$ Supply Voltage	–	–	–	–	1.7	1.95	–	–	V
Ambient Operating Temperature	–40	85	–40	85	–40	85	–30	85	°C
Load Capacitance (C_L)	30		30		30		50		pF
Output Circuit Protection Resistance (R_1)	–	–	15.3k		25k		50		Ω
Load Resistance (R_2)	–	–	11.3		25		–		kΩ
Input Rise and Fall Times		5ns		1ns/V		10ns		5ns/V	
Input Pulse Voltages	0 to V_{DDN}		0 to V_{CCS}		0 to V_{CCF}		0 to V_{CCP}		V
Input and Output Timing Ref. Voltages	0.9		$V_{CC}/2$		$V_{CC}/2$		$V_{DDQ}/2$		V
Output Timing Reference Voltage	–	–	$0.3V_{CC}$	$0.7V_{CC}$	–	–	$0.3V_{CC}$	$0.7V_{CC}$	V

Figure 5. AC Measurement I/O Waveform

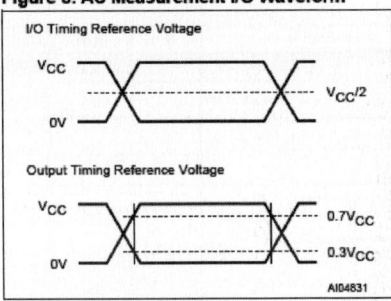

Note: V_{CC} is V_{CCF}, V_{DDN}, V_{CCP} or V_{CCS}.

Figure 6. AC Measurement Load Circuit

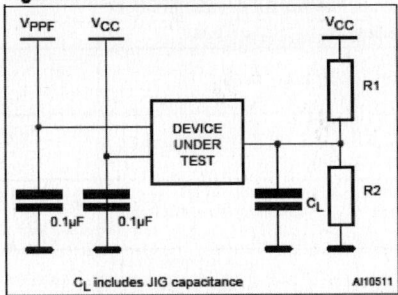

Note: V_{CC} is V_{CCF}, V_{DDN}, V_{CCP} or V_{CCS}.

Table 7. Device Capacitance

Symbol	Parameter	Test Condition	Min	Max	Unit
C_{IN}	Input Capacitance	$V_{IN} = 0V$		20	pF
C_{OUT}	Output Capacitance	$V_{OUT} = 0V$		30	pF

Note: Sampled only, not 100% tested.

Table 8. NOR Flash memory DC Characteristics

Symbol	Parameter	Test Condition		Min	Max	Unit
I_{LI}	Input Leakage Current	$0V \leq V_{IN} \leq V_{CCF}$			±1	µA
I_{LO}	Output Leakage Current	$0V \leq V_{OUT} \leq V_{CCF}$			±1	µA
I_{CC}[(2)]	Supply Current (Asynchronous Read)	$\overline{E}_F = V_{IL}, \overline{G}_F = V_{IH},$ f = 6MHz			10	mA
I_{CCB}	Supply Current (Burst Read)				30	mA
I_{CC1}	Supply Current (Standby)	$\overline{E}_F = V_{CCF} \pm 0.2V,$ $\overline{RP}_F = V_{CCF} \pm 0.2V$			100	µA
I_{CC2} [(1,2)]	Supply Current (Program/Erase)	Program/Erase Controller active	$\overline{WP}_{F/p} = V_{IL}$ or V_{IH}		20	mA
			$\overline{WP}_{F/p} = V_{PPF}$		20	mA
V_{IL}	Input Low Voltage			−0.5	0.8	V
V_{IH}	Input High Voltage			$0.7V_{CCF}$	$V_{CCF} +0.3$	V
V_{PP}	Voltage for V_{PPF} Program Acceleration	$V_{CCF} = 2.7V \pm 10\%$		11.5	12.5	V
I_{PP}	Current for V_{PPF} Program Acceleration	$V_{CCF} = 2.7V \pm 10\%$			15	mA
V_{OL}	Output Low Voltage	$I_{OL} = 1.8mA$			0.45	V
V_{OH}	Output High Voltage	$I_{OH} = -100µA$		$V_{CCF} -0.4$		V
V_{ID}	Identification Voltage			11.5	12.5	V
V_{LKO}	Program/Erase Lockout Supply Voltage			1.8	2.3	V

Table 9. PSRAM DC Characteristics

Symbol	Parameter		Test Condition		Min	Max	Unit
I_{CC}	V_{CCP} Active Current		$V_{CCP} =1.95V, V_{IN} = V_{IH}$ or $V_{IL},$ $\overline{E}1_P = V_{IL}$ and $E2_P = V_{IH},$ $I_{OUT} = 0mA$	$t_{AVAX} = $ min.		30	mA
				$t_{AVAX} = 1$ µs		5	mA
I_{CCP}	V_{CCP} Active Current (Page Read)		$V_{CCP} = 1.95V, V_{IN} = V_{IH}$ or $V_{IL},$ $\overline{E}1_P = V_{IL}$ and $E2_P = V_{IH},$ $I_{OUT} = 0mA, t_{AVAX2} = $ min.			15	mA
I_{CCB}	V_{CCP} Active Current (Burst Read)		$V_{CCP} = 1.95V, V_{IN} = V_{IH}$ or $V_{IL},$ $\overline{E}1_P = V_{IL}$ and $E2_P = V_{IH}, BL = $ Continuous, t_K[(1)]$=t_{KHKH}$ min, $I_{OUT} = 0mA$			35	mA
I_{PASR}	V_{CCP} Power Down Current	Deep Power-Down	$V_{CCP} = 1.95V, V_{IN} = V_{IH}$ or $V_{IL},$ $E2_P \leq 0.2V$			10	µA
		16Mb PASR				130	µA
		32Mb PASR				160	µA
I_{SB}	V_{CC} Standby Current		$V_{CCP} = 1.95V, V_{IN}$ (including K) $= V_{IH}$ or $V_{IL},$ $\overline{E}1_P = E2_P = V_{IH}$			1.5	mA
			$V_{CCP} = 1.95V$ V_{IN} (including K) $\leq 0.2V$ or V_{IN} (including K) $\geq V_{CCP} - 0.2V$ $\overline{E}1_P = E2_P \geq V_{CCP} - 0.2V$	$T_A \leq 85°C$		300	µA
				$T_A \leq 40°C$		130	µA
			$V_{CCP} = 1.95V, t_K$[(1)]$=t_{KHKH}$ min $V_{IN} \leq 0.2V$ or $V_{IN} \geq V_{CCP} - 0.2V$ $\overline{E}1_P = E2_P \geq V_{CCP} - 0.2V$			800	µA
I_{LI}	Input Leakage Current		$0V \leq V_{IN} \leq V_{CCP}$		−1.0	1.0	µA
I_{LO}	Output Leakage Current		$0V \leq V_{OUT} \leq V_{CCP}$ (Output Disabled)		−1.0	1.0	µA
V_{OH}	Output High Voltage		$V_{CCP} = 1.7V, I_{OH} = -0.5mA$		1.4		V
V_{OL}	Output Low Voltage		$I_{OL} = 1mA$			0.4	V

Note: 1. t_{KHKH} is the clock period.
2. All voltages are referenced to V_{SS}.
3. DC Characteristics are measured after power-up.
4. TBD stands for "to be defined".

Table 10. NAND Flash DC Characteristics

Symbol	Parameter		Test Conditions	Min	Typ	Max	Unit
I_{DD1}	Operating Current	Sequential Read	t_{RLRL} minimum $\overline{E}_N = V_{IL}$, $I_{OUT} = 0$ mA	-	8	15	mA
I_{DD2}		Program	-	-	8	15	mA
I_{DD3}		Erase	-	-	8	15	mA
I_{DD5}	Stand-By Current (CMOS) 128Mb, 256Mb, 512Mb devices		$\overline{E}_N = V_{DDN}-0.2$, $\overline{WP}_N = 0/V_{DDN}$	-	10	50	µA
	Stand-By Current (CMOS) 512Mb and 1Gb Dual Die devices			-	20	100	µA
I_{LI}	Input Leakage Current		$V_{IN} = 0$ to $V_{DDN}max$	-	-	±10	µA
I_{LO}	Output Leakage Current		$V_{OUT} = 0$ to $V_{DDN}max$	-	-	±10	µA
V_{IH}	Input High Voltage		-	$V_{DDN}-0.4$	-	$V_{DDN}+0.3$	V
V_{IL}	Input Low Voltage		-	-0.3	-	0.4	V
V_{OH}	Output High Voltage Level		$I_{OH} = -100µA$	$V_{DDN}-0.1$	-	-	V
V_{OL}	Output Low Voltage Level		$I_{OL} = 100µA$	-	-	0.1	V
I_{OL} ($R\overline{B}_N$)	Output Low Current ($R\overline{B}_N$)		$V_{OL} = 0.1V$	3	4		mA
V_{LKO}	V_{DDN} Supply Voltage (Erase and Program lockout)		-	-	-	1.5	V

Table 11. SRAM DC Characteristics

Symbol	Parameter	Test Condition	Min	Typ	Max	Unit
I_{CC1} [1,2]	Operating Supply Current	$V_{CCS} = 1.95V$, $f = 1/t_{AVAV}$, $I_{OUT} = 0mA$			12	mA
I_{CC2} [3]	Operating Supply Current	$V_{CCS} = 1.95V$, $f = 1MHz$, $I_{OUT} = 0mA$			2	mA
I_{LI}	Input Leakage Current	$0V \leq V_{IN} \leq V_{CCS}$	−1		1	µA
I_{LO} [4]	Output Leakage Current	$0V \leq V_{OUT} \leq V_{CCS}$	−1		1	µA
I_{SB} [3]	Standby Supply Current CMOS	$V_{CCS} = 1.95V$, $\overline{E1}_S \geq V_{CCS}-0.2V$ or $E2_S \leq 0.2V$ or $\overline{UB}_S = \overline{LB}_S \geq V_{CCS}-0.2V$, $f = 0$	1		20	µA
V_{IH}	Input High Voltage		1.4		$V_{CCS} + 0.4$	V
V_{IL}	Input Low Voltage		−0.5		0.4	V
V_{OH}	Output High Voltage	$I_{OH} = −100µA$	1.5			V
V_{OL}	Output Low Voltage	$I_{OL} = 100µA$			0.2	V

Note: 1. Average AC current, cycling at t_{AVAV} minimum.
2. $\overline{E1}_S = V_{IL}$, $E2_S = V_{IH}$, $\overline{UB}_S$ or/and $\overline{LB}_S = V_{IL}$, $V_{IN} = V_{IH}$ or V_{IL}.
3. $\overline{E1}_S \leq 0.2V$ or $E2_S \geq V_{CCS}-0.2V$, $\overline{LB}_S$ or/and $\overline{UB}_S \leq 0.2V$, $V_{IN} \leq 0.2V$ or $V_{IN} \geq V_{CCS}-0.2V$.
4. Output disabled.

PART NUMBERING

Table 13. Ordering Information Scheme

Example: M36 W A R 7 7 7 3 0 ZAQ T

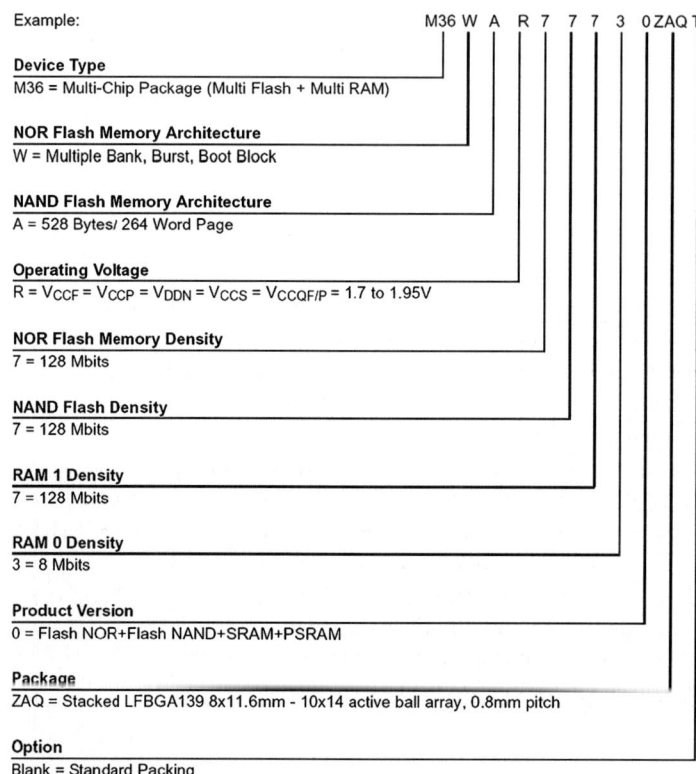

Device Type

M36 = Multi-Chip Package (Multi Flash + Multi RAM)

NOR Flash Memory Architecture

W = Multiple Bank, Burst, Boot Block

NAND Flash Memory Architecture

A = 528 Bytes/ 264 Word Page

Operating Voltage

R = V_{CCF} = V_{CCP} = V_{DDN} = V_{CCS} = $V_{CCQF/P}$ = 1.7 to 1.95V

NOR Flash Memory Density

7 = 128 Mbits

NAND Flash Density

7 = 128 Mbits

RAM 1 Density

7 = 128 Mbits

RAM 0 Density

3 = 8 Mbits

Product Version

0 = Flash NOR+Flash NAND+SRAM+PSRAM

Package

ZAQ = Stacked LFBGA139 8x11.6mm - 10x14 active ball array, 0.8mm pitch

Option

Blank = Standard Packing

T = Tape & Reel Packing

E = Lead-free Standard packing

F = Lead-free Tape & Reel packing

Devices are shipped from the factory with the memory content bits erased to '1'. For a list of available options (Speed, Package, etc.) or for further information on any aspect of this device, please contact the STMicroelectronics Sales Office nearest to you.

Appendix A

G. Campardo and R. Micheloni

The aim of this appendix is to provide further details on the internal organization of a Flash memory device in order to complete the exposition of the topics related to the nonvolatile world. The method of connecting the memory cells together into a matrix will be shown, as well as the connection to the circuitry which allows programming, erasing, and reading memory content.

Analysis of the various topics is developed on a block-by-block basis, in order to provide a hopefully comprehensive idea on the way the memory is built, together with the functionality required by its main components. The concepts can then be investigated further by means of the vast amount of existing resources.

For the sake of the exposition, various figures have been prepared in order to show the relative positions of the different blocks and their connections. Finally, the device under analysis is just an exemplification of a real one, where some blocks have been either simplified or removed for the sake of clarity.

A.1 Basic Elements

Figure A.1 is a photograph of a cross-section along a matrix column (bitline) in which the main elements of a Flash cell are shown. The control gate is composed of two layers: the polysilicon of the control gate (poly2) is below, and the silicide (used to decrease the resistivity of the row) is above. The interpolyoxide is composed of several layers as well, in order to improve its retention quality. The floating-gate, which is the distinctive element of the nonvolatile memory cell, is called poly1. Finally, Fig. A.1 also shows the junction areas of the source and the drain and the contacts between the bitline and the drain of the cell.

Drain and source contacts are shared by two adjacent cells, in order to optimize area occupation inside the matrix. This kind of configuration is referred to as NOR architecture, because of the way the cells are connected between the ground (at the source) and the bitline (at the drain). Other architectures exist, but we will concentrate on NOR in the following paragraphs. Fig. A.1 also shows the biasing voltages used in the different operating modes.

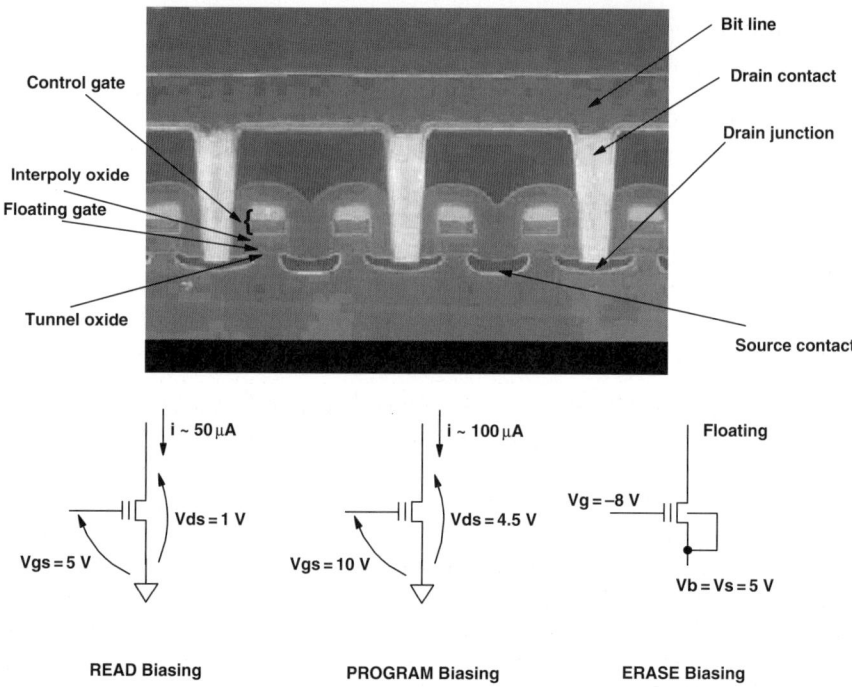

Fig. A.1 Main elements composing a Flash cell and biasing voltages used during read, program, and erase operations

A.2 Read Operation

A NOR Flash cell is said to be read in "current mode." Program and erase operations modify the threshold of a memory cell; by convention, written cells are those with an excess of electrical charge inside the floating-gate, therefore featuring a higher threshold, while the erased cells have an excess of positive charge inside the floating-gate, and thus a lower threshold. Applied voltage being equal, a written cell sinks less current than an erased one. A read operation is performed by biasing the cells while sensing the current flowing across them by measuring the related voltage drop on a resistive load (please refer to Chapter 2 for further details).

When dealing with a read operation, the designer must focus on the following targets: execution speed, read accuracy (influenced by the parasitic components—resistance toward the various terminals, unwanted voltage drops on power supplies, parasitic capacitances, and so on), and the accuracy of the voltages used. It is also important to avoid electrical stress on the unaddressed cells which share either the same wordline or the same bitline with the addressed cell.

A.3 Program Operation

In this case, the principle known as CHE (*Channel Hot Electron*) is used to inject a negative charge into the floating-gate, which modifies the threshold voltage of the cell and therefore its logical content. A current flow is enabled in the channel of the cell by applying a high voltage to both the drain and the gate of the cell; typical values for a state-of-the-art process are around 4.5 V and 10 V, respectively. The resulting current flowing in the channel is on the order of $100\,\mu A$.

The values of both the voltages and the current required by the program operation have changed over the years, mainly because of the need of lowering the operating supply voltage of the devices. In the beginning, devices featured a dual-voltage supply, namely a VDD at 5 V and a VPP pin at 12 V, dedicated to both program and erase operations. Then single-voltage supply devices were designed, and the necessary supply has been reduced from 5 V to 3 V and lately to 1.65 V; 0.9 V being already under development for the next generation of devices. Such a trend has been dictated by the widespread diffusion of portable devices like mobile phones, digital cameras, MP3 players, etc., in order to extend battery life. Programming current has been reduced as well, from the initial 1 mA to $100\,\mu A$ and below.

The main limit of CHE-based programming is the need for a current flowing in the channel of the cell: if several cells are written in parallel to reduce programming time (up to 128 cells in the latest products), then charge pumps must be used, and the area of the device grows accordingly. The most important circuit feature to take into account when dealing with program operation is the correct generation of the various voltages involved, together with their related timing. It is also mandatory to avoid any electrical stress to the unaddressed cells, in order to prevent any potential modification of the stored information. Last but not least, the state of every single cell must be verified after the program pulse.

A.4 Erase Operation

One of the reasons for the success of the Flash cell lies in the fact that it is electrically erased. Previous devices, EPROM (*Erasable Programmable Read Only Memory*), were programmed using CHE, but erasing was performed by means of ultraviolet radiation: the device had to be removed from the motherboard and placed under a UV lamp, and in about 20 minutes the floating-gates reverted to their neutral state, i.e., the logic state "1." Such a complex procedure limited the use of these devices; nevertheless such products were quite simple, since the erase operation was performed on the whole matrix at the same time. On the opposite side, EEPROM (*Electrically Erasable Programmable Read Only Memory*) memories were (and still are) available: these devices are able to program and erase on a byte basis, thanks to the Fowler-Nordheim tunneling effect. The drawback is the need of a selection transistor for each byte, which has a negative impact on area. Nowadays, the maximum size for EEPROM is 1 Mb, while Flash density can be as high as 1 Gb.

Flash memory embeds both the CHE program of the EPROM and the tunneling erase of the EEPROM, performed on a broader granularity, i.e., on a sector, which is the smallest set of cells erased in a single operation. The typical sector size is 1 Mb, which is a reasonable trade-off between the user's needs and the device size, which is proportional to both the number and the size of the sectors. Each sector is electrically isolated from the others, in order to allow individual erase and to avoid undesired electrical stress.

One of the challenges for a Flash device is the maximum number of write/erase cycles, which is usually on the order of 100,000 for each sector. Such a limit is due to the aging of the cell oxide induced by the stress, and to the performance degradation caused by the worsening of charge retention.

A.5 Main Building Blocks of a Nonvolatile Memory

Figure A.2 shows a photograph of a Flash memory device where the main building blocks are highlighted. This is a 64 Mb device featuring 64 sectors, 1 Mb each, diffused in a $0.18\,\mu m$ CMOS process, with 2 polysilicon layers and 3 metal layers. The device size is approximately $40\,mm^2$. For the sake of simplicity, Fig. A.3 shows only the previously highlighted blocks, which we are going to analyze in detail.

Let's start with the 64 sectors: each of them is 1 Mb wide and is surrounded by blocks called "*Local Row Decoder*" and "*Local Column Decoder*." In order to get the physical separation of the sectors, thus avoiding any interference during the memory operation, a technique based on hierarchical decoding has been adopted.

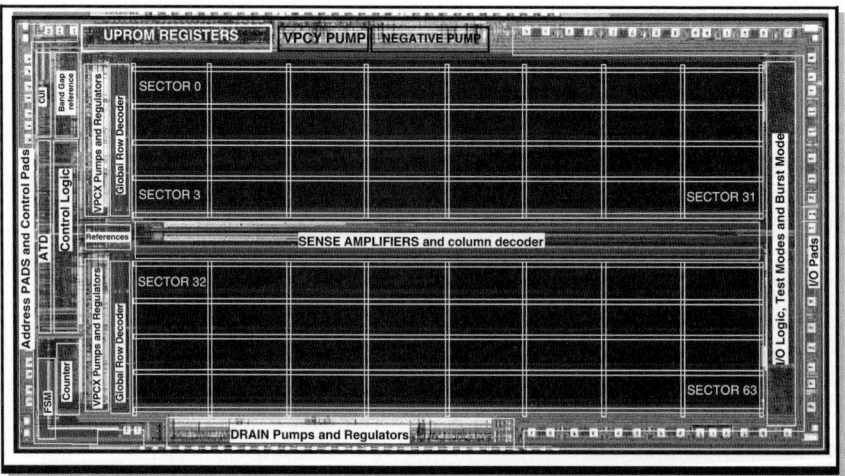

Fig. A.2 Photograph of a 64 Mb chip in which the main building blocks are highlighted

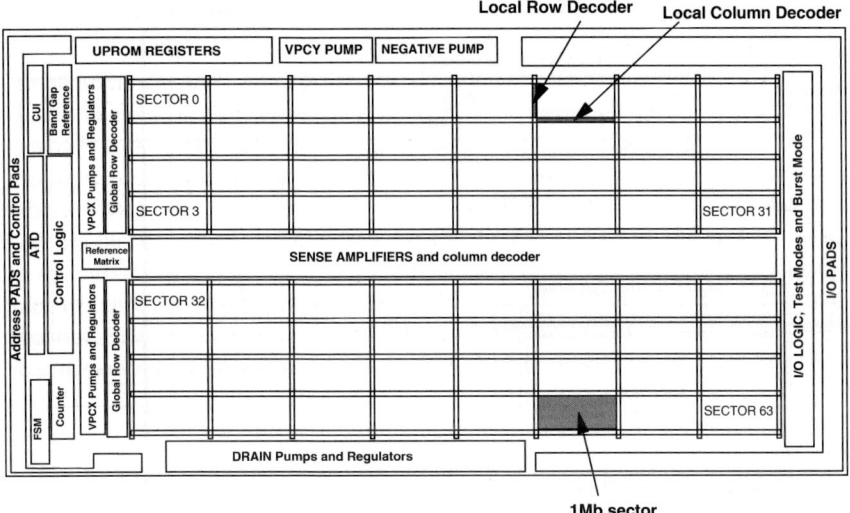

Fig. A.3 Blocks from Fig. A.2 with highlight of one sector and of row and column hierarchical decoding

Every single byte (i.e., 8 cells at the same time) or single word (i.e., 16 cells at the same time) or double word (i.e., 32 cells at the same time) is addressed by a circuit which selects the corresponding row and column in order to either read or write to the cells belonging to the same byte. The row-column crossing locates the addressed cell.

In order to avoid useless electrical stress, so-called local decoders are used, i.e., switches which are enabled to provide the required voltages exclusively to the target sector. In this way, only the cells belonging to the addressed sector are biased. In the example, the 64 sectors are separated by such hierarchical decoders, and overall cell reliability in cycling is greatly improved.

Figure A.4 shows the circuit organization of hierarchical decoders inside a memory sector. The local column decoder is composed of four pass transistors connected to the same column, called the *Main Bit Line*. Such pass transistors transfer the voltage of the global bitline (Main) to the local one. The local bitline is made of metal1, while the global bitline is made of metal3. The column decoder is divided into two parts, placed above and below the sector of the matrix; such an arrangement is due to the fact that the size of the memory cell does not allow an appropriate fit of the transistors on a single side.

Local row decoders are more complex, since they must be able to transfer several different voltages. The selected row (wordline) is biased with positive voltages both in read and in program, while a negative voltage is applied during erase. Furthermore, all the unselected lines (both in the addressed sector and in the others) must be tied to ground. All these combinations are made possible by means of a triple switch, as shown in Fig. A.4. In this case the *Main Wordline* is made of metal2,

Next to the decoders, the charge pumps which supply row decoders during read and program are located. Such pumps are built with a diode-capacitor structure, which is common practice in any Flash memory architecture. The output of the pump is regulated by a control circuit which exploits a feedback mechanism: the resulting voltage is often referred to as VPCX, and its positive values are used for read and program operations.

On the rightmost side of the device, input *Pads* corresponding to both memory addresses and control signals can be found. The main control signals for a Flash memory are the CE# (*Chip Enable*, active low), which can be used to put the device in *standby mode* (low power consumption mode—a few microamperes); OE# (*Output Enable*, active low), which is used to set the output buffers to high impedance mode, thus allowing the other devices on the board to drive the output bus; and WE# (*Write Enable*, active low), which drives the write and the command phases towards the device.

Moving from left to right, a block called ATD (*Address Transition Detector*) can be found in the middle. In order to explain the operation of this block, it is necessary to give some more detail on the Flash memory.

A memory device can be either synchronous, where an external clock is timing every function, or asynchronous, where no clock is provided and signal variations trigger any event. The former execute operations in a sequential way, using the external clock to time them. For example, in a microcontroller the instructions are executed, one after another, on every clock cycle; the disadvantage is that the time is fixed for every operation, even if some of them could be performed in a shorter time—the system speed cannot be increased.

On the other hand, asynchronous systems have no clock; every operation is executed in due time and the system speed is the maximum achievable. The challenge is to design the different operations avoiding any critical overlap of the signals due to temperature and voltage variations.

Even in the case of asynchronous nonvolatile memories, it is very useful to have a synchronization signal to warn the circuits about the beginning of a new read phase. For this purpose, the ATD circuit is designed. It is sensitive to address variations and it delivers a pulse, which is a pseudo-clock, used to trigger every read internal operation, as for instance node precharge inside the sense amplifier.

Above the ATD, the CUI (*Command User Interface*) block can be found. It is designed as either a sequencer or a finite state machine, and it is responsible for the interpretation of the command provided by the user to the device. The command set includes commands like program, sector erase, read of the *Status Register* which reports the status of the device during the various phases, program or erase suspend, etc. The possibility of suspending a modification operation allows the user access to the memory content in every moment; in this way, the system does not lose control of the memory even during the long-lasting phases of program (some microseconds) and erase (one second).

Below the ATD block, we can find the FSM (*Finite State Machine*), which is responsible for the execution of program and erase algorithms; it is a small microcontroller which can perform operations like comparisons, counts, and conditional branches. In particular, a specific block (the *"Counter"*) is used to control the duration of program and erase pulses and to count the number of such pulses; every program and every erase is composed of a predefined sequence of pulses and verifications.

Next to the CUI block, the *Band-Gap Reference* can be found. This circuit provides a voltage which is stable with respect to both temperature and power supply variations; in nonvolatile memories, all the critical voltages for program and erase are controlled by means of the Band-Gap Reference. EPROM memories allowed a drain voltage variation on the order of some hundreds of millivolts. Today, cell size imposes much better voltage accuracy, especially for drain voltage during program and for gate voltage during verify. In order to provide such accuracy, the Band-Gap Reference exploits the possibility of compensating temperature variations of the V_{BE} of a bipolar transistor with the variations of a voltage drop proportional to the thermal potential ($Vt = KT/q$), where K is Boltzmann constant, q is the charge of the electron, and T is the absolute temperature.

Now let's look at the blocks located in the upper part of the device. From left to right, the first one is the UPROM block. The fabrication of a memory device embedding millions of cells must take into account the potential defectivity, since the chance of finding one or more memory cells which do not operate correctly is quite high, because of the intrinsic complexity of fabrication processes. In order to overcome this issue, memory devices contain a set of "backup" memory cells, also known as redundancy cells, which are used as replacements for the defective cells which are found during test phases. There are specific nonvolatile registers, called UPROM, whose task is to store the addresses of the defective cells. Each time an address is selected, it is compared with the content of the UPROM; in the case of a hit, i.e., the address pertains to a defective cell, normal decoders are deactivated and a specific decoder which activates redundancy rows or columns is used. Such address translation points to a new memory location which is outside the memory space visible from outside.

Next to the UPROM registers, we can find the VPCY block, which is used to supply column decoders during program operations. In fact, it is necessary to avoid an excessive voltage drop on the pass transistors of the decoders, and acting on their driving voltage is better than increasing their size.

The last block of the upper side is the Negative Pump, which is composed of the pumps which are used to generate the negative voltages used by row decoders during erase operations. This block can also be used for the negative voltage which is needed to bias the body of the sector during the program phase in order to increase efficiency.

On the lower side of the device, the block called *Drain Pumps and Regulators* can be found. It is the charge pump which provides the positive voltage used to bias the drain of the cells during program. This charge pump provides a lower voltage than that supplied by the above mentioned VPCX, but a higher current. Furthermore, the required accuracy is different as well; therefore it is reasonable that the two pumps are different, in order to better optimize their performance.

Finally, let's analyze the two blocks located at the rightmost of the device.

The first one, called *I/O Logic, Test Modes and Burst Mode*, performs different tasks which have been grouped for the sake of simplicity. First of all, it handles the logic for input and output, which receives the data from the sense amplifiers and distributes them to the output drivers. In the state-of-the-art memory, this logic also embeds any data processing when required, like the error correction code handling.

Let's now consider the *Test Modes* topic. On top of the functionality described in the specification, there are "hidden" access modes to the memory, which can be used to carry out a thorough analysis of both the cells and the internal circuits. The access procedure is not described in the datasheet, therefore only the manufacturer knows how to do it. In this way it is possible, for instance, to apply the voltages required for program and erase from the outside, thus trying different values and finding the combination which provides the best trade-off between performance and reliability.

The last concept, *Burst Mode*, is a particular method of reading the data in a synchronous way. In general, the access time is related to the technological process used to design the device and to its complexity. As we have seen, the access time for a single, asynchronous read is around 50–100 ns. One way to speed up the read operation is to access a higher number of cells in parallel, then send a set of them to the output and, in parallel, perform a new access. In this way, after an initial latency, a higher read throughput can be sustained. Burst read requires a clock to time the different operations, and random access cannot be performed without slowing down the throughput itself.

The last block to be analyzed is the *I/O Pads*. The task of the output driver is to provide the data acquired during the read operation to the outside world. Output load is always constituted by a capacitance, whose value can vary from 30 pF to 100 pF. The structure of the driver can be likened to that of an inverter whose pull-up loads the output in case the cell is erased, or whose pull-down discharges the output in case a written cell has been read. The design of the output buffer is complicated by the fact that the charge and discharge time of the output capacitance should be as fast as possible; such a target requires high current and therefore big output transistors, thus impacting the area of the device. Furthermore, it is important to avoid any problem caused by the current used by the output drivers: disturb induced on power supplies, undesired coupling through the substrate to all the circuits inside the device, etc. In the case of 16 buffers switching at the same time, a peak current of 1 A or more can be present in the first nanoseconds.

Bibliography

1. Campardo G, Micheloni R, Novosel D (2005) VLSI-design of nonvolatile memories. Springer Series in Advanced Microelectronics
2. Lai S (1998) Flash memories: where we were and where we are going. IEDM Tech Dig Volume number:971
3. Pavan P, Bez R, Olivo P, Zanoni E (1997) Flash memory cells—an overview. Proc IEEE 85:1248

4. Prince B (1993) Semiconductor memories. A handbook of design manufacture and application. Wiley
5. Riccò B et al (1998) Nonvolatile multilevel memories for digital applications. Proc IEEE 86:2399–2421
6. Cappelletti P et al (1999) Flash memories. Kluwer, Norwell, MA
7. Dipert B, Levy M (1994) Designing with flash memory. ANNABOOKS, San Diego
8. Campardo G, Micheloni R (2003) Special issue on flash memory technology. Proc IEEE 91(4):483–488
9. Wang ST (1979) On the I-V characteristics of floating-gate MOS transistors. IEEE Trans Electron Dev ED-26(9):page numbers
10. Ohkawa M et al (1996) A 9.8mm 2 die size 3.3V 64Mb flash memory with FN-NOR type four level cell. IEEE J Solid-St Circ 31(11):1584
11. Rolandi PL et al (1998) 1M-cell 6b/cell analog flash memory for digital storage. IEEE ISSCC Dig Tech Pap Volume number:334–335
12. Micheloni R et al (2002) A 0.13-μm CMOS NOR flash memory experimental chip for 4-b/cell digital storage. Proc 28th European solid-state circuit conference (ESSCIRC), pp 131–134
13. Bauer M et al (1995) A multilevel-cell 32Mb flash memory. IEEE Int Solid-St Circ Conf Dig Tech Pap, pp 132–133

About the Authors

Pietro Baggi was born in Bergamo in 1970 and graduated from the Politecnico di Milano in Milan, Italy in 1996 with a degree in electrical engineering. Baggi began working on Sonet-SDH technology for Alcatel in Vimercate (ME) as a software designer in 1996. From 1998 to 2000, he worked for Ansaldo Industrial Systems (ME) and managed the development of firmware for PWM rectifiers. Since 2000, he has worked for STMicroelectronics and has been responsible for the design of micro-controllers for high-density memory (FLASH) cards.

Luca Benini is full professor at University of Bologna. He received a PhD degree in electrical engineering from the Stanford University in 1997. His research interests are in the design of systems for ambient intelligence, from multi-processor systems-on-chip/networks-on-chip to energy-efficient smart sensors and sensor networks. From there his research interests have spread into the field of biochips for the recognition of biological molecules, and into bioinformatics for the elaboration of the resulting information and further into more advanced algorithms for in-silicon biology. He has published more than 350 papers in peer-reviewed international journals and conferences, four books, several book chapters and two patents. He is a Fellow of the IEEE.

Davide Bertozzi is an assistant professor at University of Ferrara (Italy). He received his PhD from the University of Bologna in 2003, with a dissertation on the energy-efficient connectivity of devices at different levels of abstraction (from chip to wide-area networks). As a post-doc, he has been technical leader of several projects concerning multi-processor system-on-chip (MPSoC) design, with emphasis on disruptive interconnect technologies (networks-on-chip). He has been visiting researcher at NXP, Samsung, STMicroelectronics, NEC America and Stanford University. His current research activity addresses the intricacies of MPSoC design under severe technology constraints and of resource management for parallel integrated systems.

Roberto Bez received the doctor degree in physics from the University of Milan, Italy, in 1985. In 1987 he joined STMicroelectronics and since then he worked on the non-volatile memory technology development in the R&D department.

Flash memories. Then he worked on several projects on memory controller; he was responsible for developing MMC/SD memory controller. Since 2007 till date he is working for Qimonda on NAND Flash memories as digital design team leader.

Agostino Pirovano was born in Italy in 1973. He received the Laurea degree in electrical engineering from the Politecnico di Milano, Italy, in 1997, and the PhD degree at the Department of Electrical Engineering, Politecnico di Milano, Italy, in 2000. In 2003 he joined the Non-Volatile Memory Technology Development Group of the Advanced R&D of STMicroelectronics. Currently he is in charge of the investigation of emerging NVM technologies and he is actively working on the electrical characterization and modeling of phase-change non-volatile memories.

Giovanni Porta graduated as an electrical engineer from Pisa University. After working for Texas Instruments in Avezzano developing software, he worked for Marconi Communication as a GSM/GPRS DSP developer and for quite a while as a software product manager. He transferred to Sony-Ericsson Mobile Communication in Monaco as a software manager. Since 2004, he has worked in STM as a software development manager for Flash NOR in Arzano.

Paolo Pulici was born in 1979. He got the Laurea degree (cum laude) and PhD in electronic engineering, in 2003 and 2007, respectively. His PhD research activities covered the signal integrity analysis and the evaluation of electromagnetic interferences principally in system in package devices. From 2004 to 2007, he collaborated with STMicroelectronics in the signal integrity analysis. In 2007, he joined STMicroelectronics about BCD devices, studying in particular ADC circuitries. He was co-author of some papers and patents about SiP and signal integrity issues.

Silvia Radente graduated in chemistry in 1993 at the Public University in Milan. She collaborated until 1994 with Ronzoni Institute for the development/recycling of alternative materials coming from materials exhausted from the working of the sugar beets. She is employed in STM from 1994, earlier in the group of Corporate Package Development, and from 1999 in the quality department memories/division, where she has introduced the activity of qualification of board level for products BGA.

Roberto Ravasio was born in Carvico (BG) in 1966. He graduated at the Politecnico of Milan in 1996. From 1996 to 2000 he worked for Italtel-Siemens as a designer of radio mobile systems. From 2000 to 2006 for STMicroelectronics and since 2007 for Qimonda, he has been developing NOR and NAND multilevel Flash memories with embedded Hamming and BCH ECC. Today he is developing a 16G NAND. He is author of several patents and papers and co-author of two wonderful children, Giulia and Luca.

Giuseppe Russo has a master degree in ingegneria informatica from the University of Naples "Federico II". He started at STMicroelectronics in 2002 as software engineer in the Flash Memory Product Group (MPG). Giuseppe has worked in the development and designing of software for file and data management on NOR Flash memories. Currently he coordinates the software development team inside the wireless memory business division.

Miriam Sangalli was born in 1974 in Melzo (MI); she graduated in 2000 in electronic engineering at the "Universita' degli Studi di Pavia". For her graduation thesis she worked at STMicroelectronics in Agrate Brianza (MI) on a Flash NOR memory test chip, storing 4 bit/cell. From 2000 to 2007 she worked there as analog designer in multilevel NOR Flash memories, and then in multilevel NAND Flash memories. Since 2007 she works as analog designer in Qimonda.

Pier Paolo Stoppino was born in Milan in 1973. He graduated with a degree in electrical engineering from Politecnico di Milano in 2000. His thesis was on the problems associated with testing and assembling the first multi-memory device commercialized by STMicroelectronics. Since 2001, he has worked for STMicroelectronics in Agrate in the 3D Integration Group and is responsible for the development of the multichip.

Federico Tiziani was born in Rome, Italy, in 1969 and graduated in electronic engineering at the University of Vercelli in 1997, having written his thesis on microelectronics devices. In 1999, he joined the Memory Product Group in STMicroelectronics as firmware design engineer for Flashcard device. In 2001 he moved to the System & Application group and by 2004 he was promoted to System & Application manager for Flashcard market. Now he is in charge of the Technical Marketing team in the DATA business group.

Gian Pietro Vanalli was born in 1956. He got the Laurea degree in electronic engineering in 1986. In 1987 he joined STMicroelectronics as digital designer in the Telecom division. From 1994 to 2002 he worked as system architect in the Wireline division in telecommunications projects for network applications (broadband and narrowband). From 2002 to 2007 he worked on electric aspects in system in package devices. Recently, he joined the Data division inside the Flash Memory Group to work on new developments in signal transmission in Data applications.

Index

Printing: Krips bv, Meppel, The Netherlands
Binding: Stürtz, Würzburg, Germany